NORTHERN BEE BOOKS

Scout Bottom Farm, Mytholmroyd, West Yorkshire

www.northernbeebooks.co.uk

The Small Hive Beetle, *Aethina tumida* Murray

by

Wm. Michael Hood, PhD

ISBN: 978-1-912271-07-8

Design: SiPat.co.uk

Published by Northern Bee Books, 2017

Scout Bottom Farm
Mytholmroyd
Hebden Bridge
HX7 5JS (UK)

www.northernbeebooks.co.uk

Tel: +44 1422 882751

NOTE: front cover photo of small hive beetles, *Aethina tumida* Murray, alongside honey bees, *Apis mellifera* L. Source: Jessica Lawrence, Eurofins Agroscience Services.

The Small Hive Beetle, *Aethina tumida* Murray

Wm. Michael Hood, PhD
Professor Emeritus Entomology

Emeritus College

Clemson University, South Carolina, USA

Preface

The importance of the role played by honey bee pests in the world is becoming more recognized each year, not only because of attention given to the pest species, such as the varroa mite and small hive beetle, but also because of the increasing realization that honey bees are extremely valuable to nature and humanity.

The purpose of this book is to familiarize the beekeeping public with a new honey bee pest, the small hive beetle, *Aethina tumida* Murray. The small hive beetle is native to Sub-Saharan Africa where it is known as a minor honey bee pest. Since 1996, the small hive beetle has become an invasive pest as introductions have been recorded in five continents other than its native Africa including Asia, Australia, Europe, North America, and South America. Additional introductions to other regions of the world and further spread of the beetle in newly invaded countries are expected.

In addition to briefly discussing other beetle pests found in honey bee colonies, I will present an overview of the small hive beetle including its threat to other social bees such as bumble bees and stingless bees. This book provides the beekeeper fundamental and important information about small hive beetle management. Topics covered include an historical review, biology, diagnosis, genetic diversity, economic importance, and a review of the current management recommendations of the small hive beetle including: acceptable pest levels, preventive measures, monitoring practices, genetic control, mechanical control, physical control, biological control, and chemical control. In the United States and Australia, small hive beetles are well known for their destructive effects on honey bee colonies, especially in the warmer regions of the countries. The primary scope of this book is to introduce bee-keepers to this latest honey pest to enter many new regions of the world and to discuss practical, safe, and sustainable means of control.

Acknowledgments

There are a number of individuals who I would like to thank for their generous support in preparing this book. R. Bellinger, C. Bryan, C. Cervancia, K. Delaplane, A. Dupo, L. de Guzman, J. Ellis, K. Flottum, A. Frake, T. Gretchen, S. Hinkley, J. Lawrence, C. Linder, A. Locsin, J. Lotz, G. Levot, M. Nolan, B. Oldroyd, S. Peterson, E. Polintan, J. Tew, K. Walker, N. Wright, and J. Zawislak provided many excellent diagrams, graphs, and photos for this book.

Contents

Chapter 1

Other beetle species found in honey bee colonies

Since the common name "small hive beetle" refers to the species *Aethina tumida*, this leaves one to wonder if there is also a large hive beetle? Yes, indeed there are two related species of large hive beetles: *Oplostomus fuligineus* (Olivier) and *Oplostomus haroldi* (Whitte). Large hive beetles are in the scarab beetle family (Scarabaeidae) whereas the unrelated small hive beetle is found in the sap-feeding beetle family (Nitidulidae). Although small hive beetles are currently well known throughout many parts of world, the large hive beetles also native to Africa, have not been introduced to other continents (http://beebugs.blogspot.com/2013/05/since-its-called-small-hive-beetle-does.html).

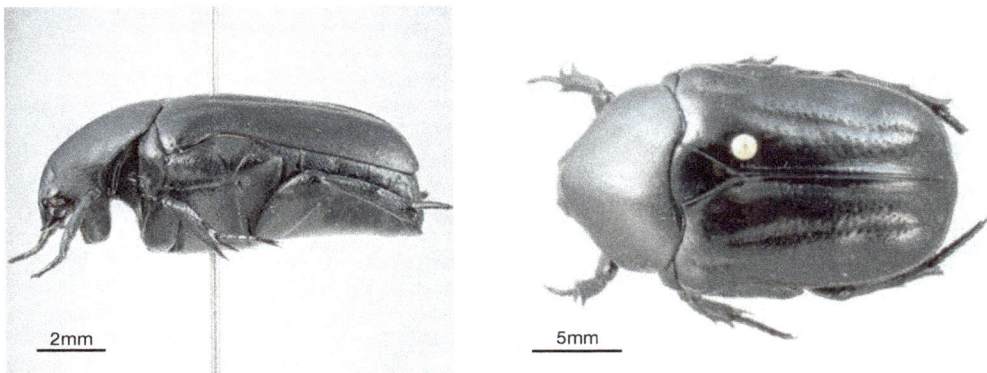

Fig. 1. Lateral and Dorsal Views of the Large Hive Beetle (Oplostomus fuligineus) Adult. Source: Simon Hinkley and Ken Walker, Museum Victoria.

Fig. 2 Large Hive Beetle Adults on Honey Bee Brood Comb. Source: Ben Oldroyd, University of Sydney, Australia.

Life Cycle

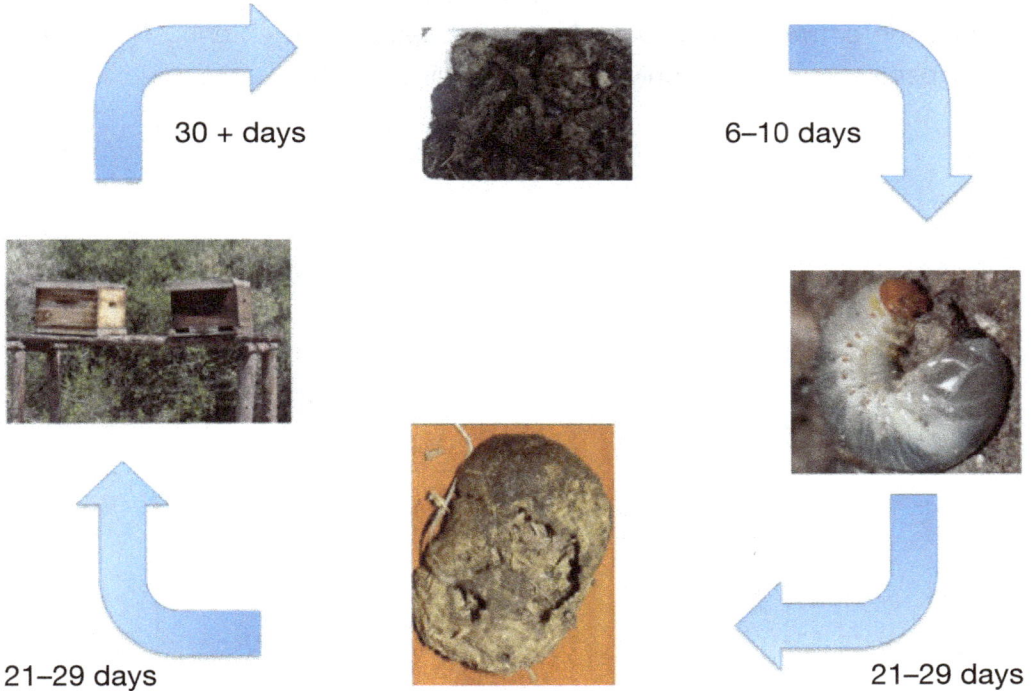

30 + days

6–10 days

21–29 days

21–29 days

Fig. 3. Life Cycle of the Large Hive Beetle. Source: Ben Oldroyd, University of Sydney.

However, it is recommended that all beekeepers become familiar with the appearance of large hive beetles and report their occurrence immediately to apiary regulatory authorities, if found outside their native range in Africa.

The Large hive beetle (*Oplostomus fuligineus*) adult is 20-23 mm (0.8 - 0.9 inches) in length and is shiny black and glabrous, antennal club 3-lamellate and pronotum coarsely punctate. The beetle spends about 30 days of its approximately 87-99+ days life cycle inside honey bee colonies in Africa. The adults feed mostly on honey bee brood instead of honey and pollen and can quickly destroy much of the brood and comb within a bee colony. It appears the large hive beetles exit the bee colony and must find decomposing plant material and herbivore (commonly cattle dung) for oviposition and development (Donaldson 1989). There seems to be no defense of these large beetles once they enter a bee colony, therefore beekeepers in Africa reduce the entrance size to exclude the beetles from entering their hives. http://beeaware.org.au/archive-pest/large-hive-beetle/

Chapter 2

History of the small hive beetle

The small hive beetle (*Aethina tumida* Murray) is a minor honey bee pest found in many African countries. Small hive beetles were first collected from Old Calabar on the west coast of Africa and sent to Andrew Murray in London in 1867 for identification (Murray 1867). The beetle is native to sub-Saharan Africa where it is known as a scavenger in honey bee colonies (Hepburn and Radloff 1998) and received little attention prior to its discovery in the USA in 1998. African honey bee subspecies apparently possess behavioral traits which allow them to prevent small hive beetle depredation.

In their native range, the beetles are better known for their damaging activities around honey houses and their effect on weak or stressed honey bee colonies. The small hive beetle has now become established in new regions of the world where the pest has become a major problem for many beekeeping operations. Apparently European honey bees (*Apis mellifera* sp.) lack some of the behavioral traits of African bees in tolerating the pest and results in increased colony losses. Even strong honey bee colonies have succumbed to the effects of small beetles and died in heavily-beetle-infested areas.

Prior to 1998, only two significant research investigations were conducted and published on small hive beetles (Lundie 1940 and Schmolke 1974). Their work covered several aspects of the biology and control of this hive pest in its native range. In Africa, the beetle is considered less important than the greater wax moth (*Galleria mellonella*) and the lesser wax moth (*Achroia grisella*) which are known also for their function of eliminating or cleaning up dead or weakened honey bee colonies.

Damage from small hive beetles is more apparent in honey bee colonies in the newly established areas of the world. The beetle adults and larvae feed on honey, pollen and bee brood (Hepburn and Radloff 1998). Another detrimental effect left behind by beetles is the spoilage of stored honey that probably results from beetle defecation (Lundie 1940). The fermented honey left behind in dead colonies is rejected by honey bees (which will not consume it) and is unmarketable by the beekeeper. Small hive beetles are also recognized for creating pest problems for beekeepers in and around honey houses in the new world (Hood 2004).

As of the publication of this book in 2017, small hive beetles now have a wide known presence being found in six of the seven recognized continents of the world and including several African countries along with Australia, Brazil, Canada, Costa Rico, Cuba, Egypt, El Salvador, Italy, Jamaica, Mexico, Nicaragua, Philippines and the USA, including the Hawaiian Islands.

North America

United States. The first known discovery of small hive beetles, *Aethina tumida* Murray, in North America occurred in 1996 in the city of Charleston, South Carolina, USA (Hood 1999). This first collection was made approximately 2 miles from the Port of Charleston which is a major eastern US seaport that receives worldwide goods and products (Hood 2000). Beetle specimens were collected by a small-scale beekeeper from a managed honey bee colony that had been established from a hanging swarm of bees captured from a tree located in the city of Charleston in the summer of 1996. Later in the summer and the fall, the beekeeper noticed a few small black beetles inside the hive. He collected a sample of the adult beetles and forwarded them to the Department of Entomology, Clemson University, Clemson, South Carolina, USA for identification in fall of 1996. An insect taxonomist attempted to identify the beetles, but he could only identify them to family, Nitidulidae, because keys were not available to further identify them to genus and species. Another sample of the beetles was submitted from the coastal area of South Carolina to Clemson University in 1997, but university taxonomists were again unable to identify them to genus and species. The unidentified beetle specimens were stored in the Clemson University Department of Entomology Museum (Hood 2004). During this 2 year lapse in proper identification of this unknown honey bee pest, the beetles were quickly spreading in coastal South Carolina.

In 1998, an apiary of managed honey bee colonies near St. Lucie, Florida, USA, was decimated by beetles that were properly identified by M.C. Thomas of the Florida Department of Agriculture (Elzen et al. 1999 and Sanford 1999). The adult beetle specimens collected earlier in South Carolina in 1996 and 1997 were later confirmed as *Aethina tumida*. This was the first time small hive beetles had been discovered on any continent outside of Africa.

Extensive area-wide honey bee colony inspections were made in Florida, Georgia, and South Carolina to survey for further spread following the initial identifications in June and July 1998. Honey bee colonies in several coastal counties were found to be beetle-infested in each of the three states by autumn 1998 (Hood 2000). In November 1998, small hive beetles were discovered in Scotland County, North Carolina, USA (Hopkins 1999).

The likelihood of natural spread from a single port city in the US such as Charleston to become widely spread to coastal areas of Georgia, Florida, North Carolina, and South Carolina in a 2 year period is doubtful. Beekeepers in the Charleston area are hobbyist and rarely move bees from the port city (Hood 2004). Anecdotal reports from beekeepers in the Savannah, Georgia, USA area indicated that small hive beetles occurred also in their apiaries prior to 1998. Savannah is another major port city that is located approximately 135 km (84 miles) south of Charleston. One suggestion is that small hive beetles probably arrived in the Charleston port and other ports along the southeastern US by human-assisted movement at approximately the same time aboard cargo ships loaded with a common commodity that supported the small hive beetles voyage from Africa (Hood 2004). Another unsubstantiated suggestion is that small hive beetles entered the US by "hitchhiking" with honey bee swarms or with illegally imported queens from Africa.

Regardless, small hive beetles quickly spread in the US and by December 1999, small hive beetles had been discovered in 12 states including Florida, Georgia, Iowa, Maine, Massachusetts, Minnesota, New Jersey, North Carolina, Ohio, Pennsylvania, South Carolina, and Wisconsin (Hood 2000).

Major honey bee colony losses occurred in many of the initial small hive beetle infested southeastern USA states. Six months lapsed between the time that the first beetle identifications were made and the time that an emergency use small hive beetle control product was available to the beekeeping industry (Hood 2004). Beekeepers simply had no clue how to control this new honey bee pest which allowed the small hive beetle to thrive in a new region of the world where conditions were favorable for its reproduction. During this period, affected beekeepers became frustrated with the inability to control this hive pest. In the state of Florida, a quarantine on movement of honey bee colonies was established in June 1998, but it was soon withdrawn by the Florida Department of Agriculture and Consumer Services (Fore 1998).

Small hive beetles spread as a result of natural range expansion and migratory movement of infested managed honey bee colonies, package bees, queen cages, and empty beekeeping equipment (Hood 2000). The primary means of dispersal of the beetles in such a brief period of 2 years in the US can be attributed to migratory movement of honey bee colonies annually from Florida and other southeastern states to the northeastern states for pollination purposes in the spring (Pettis et al. 2014). Pettis and Shiminuki (2000) reported that adult small hive beetles can survive for 5 days without food or water; therefore beetles may have been transported long distances also by air or vehicle on materials free of bees or beekeeping equipment. Small hive beetles have spread now to likely every state in the country.

Small hive beetles were discovered in the Hawaiian Islands in 2012. The beetles were   confirmed first on the Big Island of Hawaii where they were collected by a beekeeper in Pana'ewa (Robson 2012). The beetles became well established quickly on the Big Island, given the Hawaiian warm and humid climate and have spread now to other islands: Oahu (2010), Moloki (2011), Kanai (2012) and Maui (2011). The abundance of unmanaged feral European derived honey bee colonies has likely played a major role in the establishment of small hive beetles on the Hawaiian Islands. The tropical climate and other environmental conditions such as favorable soil habitat suitable for beetle pupation likely contributed to the massive number of beetles that naturally spread quickly throughout the islands (Conner 2011).

Canada. In August 2002, small hive beetles were discovered in Manitoba, Canada (Dixon and Lafreniere 2002). Extensive surveys were conducted in the Manitoba area in spring and summer of 2003; no further beetle activity was discovered. Reports indicated that the beetles arrived in Manitoba aboard a shipment of beeswax imported from Macgregor Wax Works, Hull, Texas, USA. An agreement was reached with the Manitoba Beekeeper's Association and the local wax rendering plant to contain the pest so that accidental introductions could be avoided in the future.

In 2010, the Ontario Ministry of Agriculture, Food, and Rural Affairs published a report indicating small hive beetle activity in Manitoba (2002 and 2006), Alberta (2006), Quebec (2008 and 2009) and Ontario (2010). The British Columbia Ministry of Agriculture published Apiculture Bulletin # 219 indicating that Provincial Apiarist Paul Kozak reported three new positive small hive beetle locations in Ontario in May 2016. Neumann et al. (2016b) noted that Ontario appears to have an established beetle population in Essex County which is located in the most southern point of Ontario along the US/Canadian border. No movement of honey bee colonies or apicultural equipment is allowed to leave that region of Ontario. When small hive beetle infested bee colonies are found in other regions of the province, the colonies are immediately exterminated or moved inside the quarantined area of the small hive beetle positive Essex County area (Dubuc 2013; Neumann et al. 2016b). The small hive beetle may not become established in some of the areas of Canada that are listed above except the area noted in Ontario. Neumann et al. (2016b) suspects that small hive beetles may not be able to become well established in many regions of Canada due to unfavorable climatic and environmental conditions.

Cuba. Small hive beetles were confirmed to be found in Cuba in 2012 (Milian 2012; Darius 2014). Darius (2014) reported the likelihood of the presence of beetles in the entire country, however he noted that official reports indicate beetles to be found in several provinces: Artemisa, Cienfuegos, La Havana, Mayabeque, Matanzas, Pinar del Rio, and Villa Clara. Although beetles have now spread to several provinces in Cuba, no reports of major effects on local honey bee colonies have been recorded (Borroto et al. 2014) which may be due to low initial infestation rates (Spiewok et al. 2007).

Jamaica. The small hive beetle was first detected in Jamaica in 2005 (FERA 2010) and reports indicate that *Aethina tumida* and *Varroa destructor* have been identified across all parishes in the country (Smith 2012). No clear evidence has been found as to how the beetles were introduced into the country. The well established beetle populations have yet to result in a serious problem for local Jamaican beekeepers who are relying mostly on their honey bees natural abilities to control the pest populations. Reports note that beekeepers are doing little to control the pest other than placement of their bee-hives on concrete surfaces (Neumann et al. 2016b). One explanation for the beetles reported little effect on the Jamaican beekeeping industry is the possible presence of Africanized bees on the island, which may allow the bees to better cope with the small hive beetles (Neumann et al. 2016b).

Australia

In October 2002, small hive beetles were discovered in honey bee colonies in Australia in Richmond, north-west of Sydney, New South Wales (Somerville 2003), which is also a coastal area. Gillespie et al. (2003) reported that the beetles were likely to have been imported into New South Wales at least 6 months prior to their official discovery. The Australian government made the decision not to try and eradicate the small hive beetle because beetles had also been found in feral honey bee colonies. An extensive survey was conducted in managed honey bee colonies from October 2002 to January 2003 in the New South Wales area and reported 120 positive detections out of more than 1,000 samples received. Small hive beetles were found mainly west of the Sydney basin around Richmond and parts of the lower Blue Mountains (Gillespie et al. 2003). Since this time, small hive beetles have also been found in Victoria and Western Australia on the north east border with Northern Territory (Learner et al. 2015). Beetle identifications were made from bee colonies in four regions, including the Sydney Basin, Cowra, Binalong, and Stroud (The Australian Beekeeper 2002). Along with discovery of beetles in managed bee colonies, 13 feral honey bee colonies in the Sydney area were confirmed to have small hive beetles (Somerville 2003).

In Australia, about 4 years passed (2002-2006) before heavy losses occurred in honey bee colonies, whereas in the US only 2 years passed (1996-1998) before heavy losses were reported (Neumann et al. 2016b). However, Neumann et al. (2016b) suspected this remarkable difference in time delay in Australia was a result of an historic drought in the country during that period of time (Horridge et al. 2005).

Europe

Portugal. In 2004, small hive beetles were intercepted and confirmed in a shipment of queens imported from Texas, USA, into Portugal (Murilhas 2004). Beetle larvae were discovered in queen cages and rigorous sanitation measures were immediately taken to prevent the introduction of beetles into the country. All bees and beetles in the shipment were killed and all colonies in the destination apiary were destroyed. The soil in the destination apiary was treated with insecticide (Murilhas 2004; Neumann and Ellis 2008; Valerio da Silva 2014). The failure of the small hive beetle to become established in Portugal was likely a result of early detection and the fact that only a few beetle larvae were imported (Neumann et al. 2016b) and no beetle adults were involved.

In 2010, European countries considered the experiences of small hive beetles in the USA and Canada and planned accordingly (Williams 2015). For example, the UK conducted a formal analysis 3 to investigate how the beetle might enter the UK. Key risks of the report included:

- "Movement of honey bees: queens and package bee (workers) for the purposes of trade

- Movement of alternative hosts e.g. bumble bees for pollination purposes

- Trade in hive products – e.g. raw beeswax and honey in drums

- Soil or compost associated with the plant trade

- Fruit imports – in particular avocado, bananas, grapes, grapefruit, kiwi, apples, mango, melons and pineapples – small hive beetles may oviposit (lay eggs) on fruit

- Movement on beekeeping clothing equipment

- Movement in freight containers and transport vehicles themselves

- Natural spread of the pest itself by flight, on its own or possibly in association with a host swarm" (Williams 2015).

The report noted that the UK has not allowed the import of bee colonies or package bees from Third Countries (outside the EU) for several years. With the exception of New Zealand, EU legislation prohibits the importation of package bees or colonies from Third Countries (Williams 2015).

Italy. In September 2014 during a survey, small hive beetles were discovered and confirmed inside three *Apis mellifera ligustica* colonies in Calabria, Italy (Palmeri et al. 2015) which is a region known for its warm Mediterranean climate. Beetle larvae and adults were found in the colonies which had been placed in a citrus orchard located in Soverto (Gioia Tauro, Reggio Calabria) in spring 2014. Colonies had been surveyed monthly by checking each frame with no signs of small hive beetles until September 5th (Palmeri et al. 2015). The colonies were immediately isolated by placement in plastic bags and killed using ethyl acetate. Source of the beetles was unknown, however small hive beetles likely arrived through the nearby port of Gioia Tauro where goods are imported from all over the world by commercial trades or by hive movement (Palmeri et al. 2015). Unfortuately, large numbers of hives move through the region from April (citrus bloom) to July (kiwi fruit pollination service) (Palmeri et al. 2105). This may have contributed toward the spread of small hive beetles in the region. Further surveys confirmed 83 more small hive beetle-infested apiaries in the Calabria area, one apiary in Sicily, and one feral colony in the municipality of Gioia Tauro (3 November 2014, Istituto Zooprofilattico Sperimentale delle Venezie 2015; Neumann et al. 2016b). The positive beetle find on the island of Sicily occurred on 7 November 2014 in Syracuse which is less than 75 miles (120 km) from Calabria (Quigley 2015a). These positive finds of beetles as a whole indicate that beetles arrived in the area well before the initial detection on 5 September 2014 (Neumann et al. 2016b).

Calabria is a region fertile with many beekeepers who migrate their colonies to fulfill pollination contracts as well as being a region where bee breeders who supplied queens and package bees to beekeepers in other European countries (Quigley 2015b). In the past, Italy has been used as a good overwintering area for Swiss, German, and Austrian beekeepers to avoid winter losses and to build strong colonies for spring. Calabria is one of the favorite overwintering grounds for beekeepers because of its good pollen and nectar flows in winter (Quigley 2015b).

On 12 December 2014, the European Union banned movement of any honey bees, bumble bees, bee equipment, and hive products from the Calabria region and Sicily to anywhere in the European Union or the European Economic Area. The ban was planned to expire on 31 May 2015 when it would be reviewed (Quigley 2015b). Italian Authorities have implemented a range of means to eradicate the pest. Several hundred bee colonies have been destroyed in beetle-infested apiaries in Italy by Italian Authorities in an effort to eradicate the pest. The chance of eradication of this hive pest is doubtful given that so many apiaries and some feral honey bee colonies (three beetle-infested wild colonies have been found and destroyed) were beetle-infested in this region, which is well known for its mild Mediterranean climate (Quigley 2015a).

Central America

Mexico. Many regions of Central America, South America, and the Caribbean appear to have favorable climatic and environmental conditions favorable to small hive beetles. Del Valle Molina (2007) first reported the discovery of small hive beetles in Mexico. The beetles are now well established in at least eight states in Mexico (Neumann et al. 2016b). Loza et al. (2014) reported extremely high adult beetle infestations in honey bee colonies in single colonies in tropical states like Yucatan. Brown and Learner (2016) noted that beetle infestations of hundreds or even thousands of adult beetles have been found in single beehives in the tropical states of Mexico. This is very surprising given that local honey bees in the region are Africanized and thought to be less susceptible to small hive beetles (Neumann et al. 2016b).

Costa Rica. The first occurrence of small hive beetles in Costa Rica was recorded in LaGarita, Santa Cecilia, La Cruz, GUANACASTE and an official report was submitted to OIE (World Animal Health Organization) on 24 August 2015 (Hernandez 2015). The discovery of beetles was made in a sentinel apiary (three hives) strategically located near the northern border with epidemiological surveillance purposes. The origin of beetles was unknown or inconclusive. The beetles were detected near the border with Nicaragua which reported discovery of beetles in 2014 (Hernandez 2015).

El Salvador. Small hive beetles were first reported in El Salvador in 2013 (Arias 2014), however a follow up survey (only 68 of 1,700 hives positive) indicated a rather localized outbreak (Neumann et al. 2016b).

Nicaragua. Small hive beetles were first reported in Rivas, Nicaragua in February 2014 (Gutierrez 2014; Calderon Fallas et al. 2015). An apiary located in San Juan del Sur of 18 honey bee colonies was affected by an infestation of small hive beetles. Out of the 18 colonies, 17 were positive for adult beetles as well as two beetle larvae were collected. Beekeepers noted that their colonies were weaker than normal and they suspected the presence of the beetles (Gutierrez 2014). In 2016, a second infestation of small hive beetles was reported in Lechecuagos, Valle Los Urroz, Leon (Diaz 2016). In the affected apiary, the honey bee colonies were noted to be strong with low beetle populations. Although the location of the small hive beetle discovery offers ideal sandy soil conditions for the beetle to reproduce, no beetle larvae or pupae were discovered during the inspections (Diaz 2016). The origin of the infestation was unknown or inconclusive. Movement control within country was one of the control measures taken (Diaz 2016).

Asia

Philippines. The first detection of small hive beetles in Asia occurred in managed *Apis mellifera* colonies in Lupon, Philippines in June 2014 (Brion 2015). It is unknown how the beetles were introduced into the country, however queen bee shipments from other countries is the suspected route (Cervancia et al. 2016). Several of the initial beetle-infested European honey bee colonies became severely infested and the majority of the colonies collapsed and died (Brion 2015). Small hive beetle-free *Apis cerana* and stingless bee (*Tetragonula biroi*) colonies were moved into the beetle-infested areas. The *A. cerana* colonies became beetle infested within 1 day and absconded within a week, whereas the stingless bees remained strong and beetle-free (Cervancia et al. 2016).

The University of the Philippines Las Ban~os began a Quick Response Service in November 2014 to interview individual beekeepers and examine their colonies for small hive beetles (Cervancia et al. 2016). Apiaries across the Philippines including 1,869 colonies in 18 provinces were inspected. All colonies in apiaries having five or less colonies were inspected. Apiaries with more than five colonies required only the weakest (6-10) colonies to be inspected. All frames and bottom boards were closely inspected.

Survey results (Cervancia et al. 2016) reported that small hive beetles are so far confined to Mindanao in the southern Philippines and no quarantine has been established to prohibit inter-island movement of bee colonies. Beekeepers were encouraged to not move colonies and other beekeeping equipment and supplies in the country to prevent further invasion of the small hive beetles. This first report of small hive beetles in Asia is troubling because this region of the world offers warm and humid conditions that are favorable to beetle development and reproduction (de Guzman and Frake 2007). Small hive beetles were not found in feral honey bee colonies and the potential impact on native *Apis cerana* and stingless bee colonies is unknown at this time.

South America

Brazil. The first record of discovery of small hive beetles in South America occurred in Brazil in March 2015 (Teixeira et al. 2016; Toufailia et al. 2017). Adult beetles were collected in an apiary of nine colonies of *Apis mellifera scutellata* in Piracicaba, Sao Paulo State, Brazil. The adult beetles were detected in an apiary of the Laboratory of Useful Insects, Department of Entomology and Acarology, University of Sao Paulo (ESALQ) (Toufailia et al. 2017). Epidemiological investigations are underway within a 20 km radius of where the first discovery of small hive beetles was made (Brown and Learner 2016).

Egypt

Small hive beetles were first detected in Egypt in Itay-Al-Baroud and subsequently in honey bee colonies along the Nile Delta (Neumann and Elzen 2004). However, more extensive surveys were conducted later in the country that could not confirm beetle activity (Hassan and Neumann 2008) and no beetle damage has been reported by local beekeepers, so it appears the small hive beetle is not well established in Egypt (Neumann et al. 2016b).

Chapter 3
Biology

The small hive beetle is in the Coleopteran family Nitidulidae which there are about 2,500 described species found globally (Habeck 2002) and 183 described species in North America (Borror et al. 1989). Most species of this family are found where plant fluids are fermenting or souring such as around decaying fruits, melons, flowing sap, or fungi. Many nitidulids are pests of fruit and stored food, and some like the small hive beetle have a close association with social hymenoptera such as bees, wasps and ants.

Laboratory diagnosis of small hive beetles is based on morphological criteria (Neumann et al. 2013) or by molecular identification (Evans et al. 2000, 2008; Lounsberry et al. 2010; Ward et al. 2007) Adult beetles average 5.7 mm (7/32 inch) in length and 3.2 mm (1/8 inch) in width. Key identification features of adult beetles are "club-shaped" antennae and elytra, which appear smaller than the abdomen, exposing the end of the abdomen (Neumann et al. 2016b). Adult beetles vary in size which is likely dependent on food quality and climate. Adult female beetles slightly outnumber and are heavier than adult males in local populations as reported by a two-state survey conducted in the southeastern US (Ellis et al. 2002c). Adults can be sexed by depressing the posterior of the abdomen exposing the genitalia. The female ovipositor is long and extends straight from the hind tip of the abdomen. The male genitalia extend at a right angle when viewed sideways ventrally (de Guzman et al. 2017).

Adult small hive beetles are strong fliers and are capable of flying several kilometers which aids in their natural spread (≥10 km (6.2 miles) being pos-

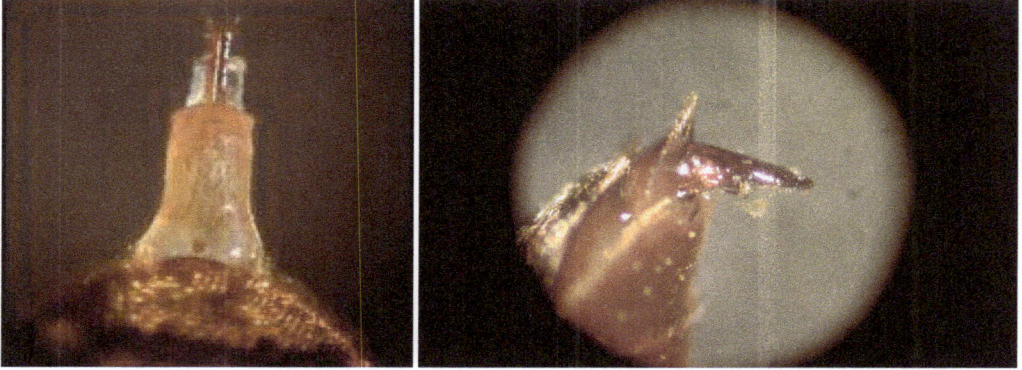

Fig. 4. Small hive beetle adult genitalia: female on left and male on right. Source: Lilia de Guzman, USDA ARS, Baton Rouge, LA.

Fig. 5. Dorsal view of small hive beetle adult. Notice clubbed antennae. Source: Jamie Ellis, University of Florida, Gainesville, FL

Fig. 6. Dorsal view of small hive beetle adult with appendages pulled underneath body. Source: Natasha Wright, Florida Dept. Agriculture and Consumer Services, Bugwood.org.

sible, Neumann and Elzen 2004). Beetles fly before or after dusk and males have been reported to fly earlier than females in the US. However, Spiewok and Neumann (2012) reported that there is a bias towards females beetles flying earlier than males in Australia and Africa. In Australia, Annand (2011) reported that the greatest number of beetle adults to enter hives occurred 2 hours before nightfall.

Fig. 7. Lateral view of a small hive beetle adult. Source: Pest and Diseases Image Library. Source: Bugwood.org

Fig. 8. Ventral side view of a small hive beetle adult. Source: J. Ellis

Small hive beetle adults are thought to be attracted to honey bee colony odors especially the honey bee alarm pheromone, but they may also be attracted to beetle pheromones which have not been identified. In olfactometric and flight-tunnel bioassays, adult small hive beetles were found to be attracted to volatiles from adult worker bees, freshly collected pollen, unripe honey and slumgum (Suazo et al. 2003).

Small hive beetles are sexually mature at about 1 week following emergence from the soil. Males mount females and copulation occurs with both males and females often mating multiple times. Small hive beetles copulate only in aggregations and sexual behavior reaches a peak at about the age of 2 to 3 weeks (Mustafa 2015). Adult females will oviposit directly on pollen or brood comb if unhindered by worker bees. It has been estimated that female beetles may potentially lay up to 1,000 eggs in their 4-6 month lifetime, although other estimates range up to 2,000 eggs. Female beetles have been reported to have a shorter adult lifetime when laying eggs on a daily basis (Somerville 2003). Mass reproduction of small hive beetles was confirmed in a research project when 80 parental beetle adults produced approximately 36,000 adult offspring in 63 days (Murrle and Neumann 2004). Other research conducted by Conklin (2012) reported initial data indicated that small hive beetles may require an extended mating period in order to be fertile which may offer an opportunity for beetle control by mating disruption strategies. Somerville

Fig. 9. Massive Number of Small Hive Beetle Adults Aggregating On Hive Bottom. Source: Charles Linder

(2003) reported that small hive beetle adults live 1-12 months and can survive up to 16 months in laboratories.

In a research project which 5-frame honey bee colonies were inundated with small hive beetles, female beetles were observed chewing holes in capped bee brood and ovipositing eggs on bee pupae (Ellis 2004). In addition, adult beetles were reported to oviposit in capped bee brood through slits they chewed in the side of adjacent empty cells. Neumann and Hartel (2004) noted that small hive beetle eggs are often eaten by worker bees in both under-neath cell cappings (Ellis et al. 2003a) or in gaps and other unprotected areas in the hive.

Small hive beetle eggs are normally laid in clusters of 10-30 plus (Stedman 2006) and are pearly white in color; the eggs are about 1.4 mm (1/16 inch) long and 0.26 mm (1/64 inch) wide. Female beetles lay eggs in cracks and

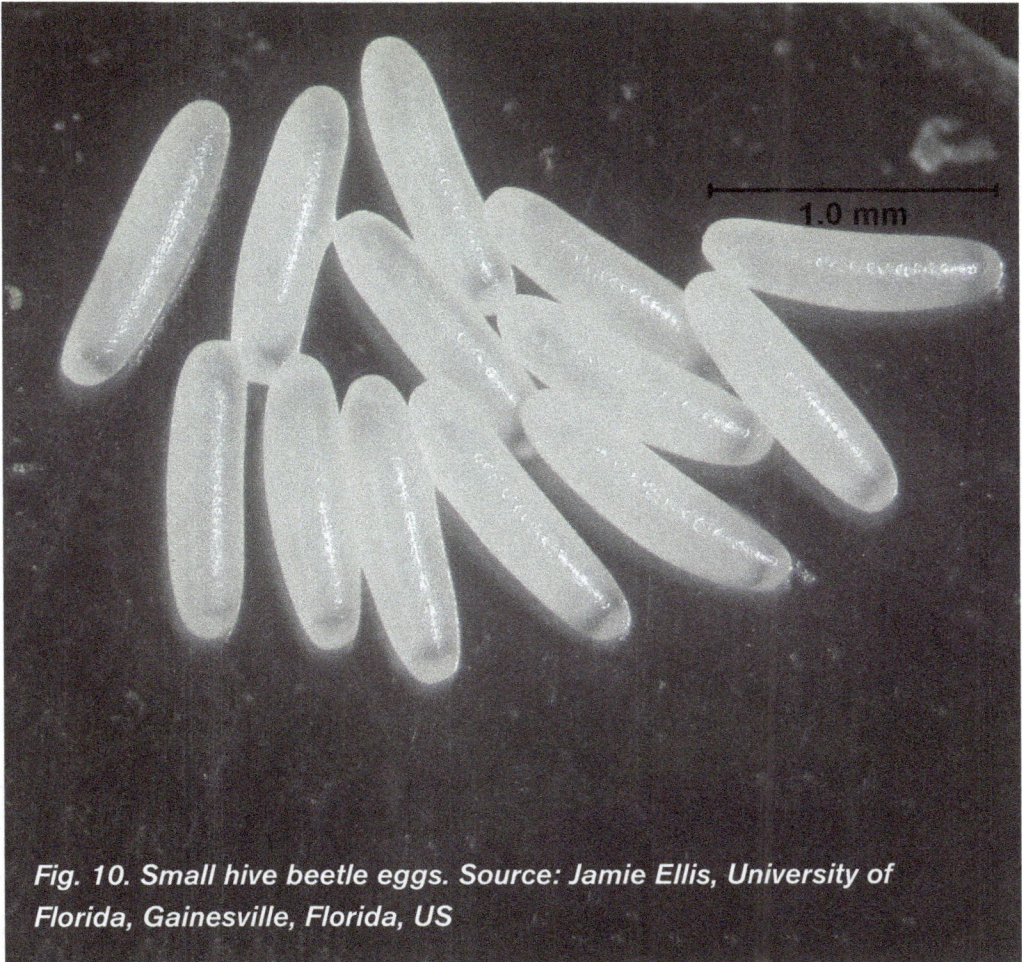

Fig. 10. Small hive beetle eggs. Source: Jamie Ellis, University of Florida, Gainesville, Florida, US

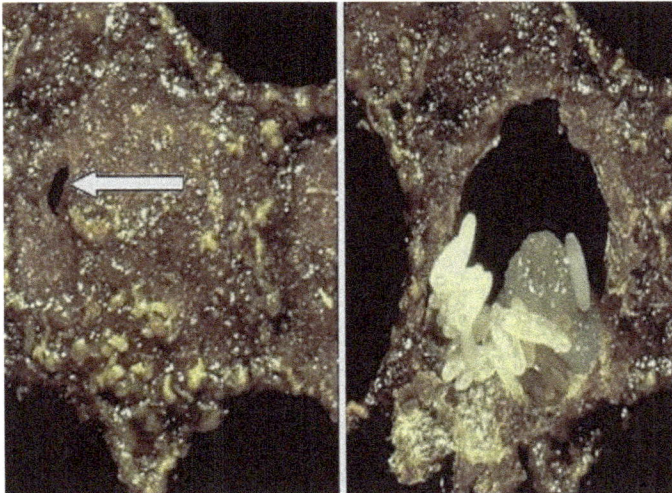

Fig. 11. White arrow points to hole in capping that female beetle ovipositor entered to lay eggs on honey bee pre-pupa as shown on right when capping has been removed. Source: Keith Delaplane, University of Georgia, USA. Bugwood.org.

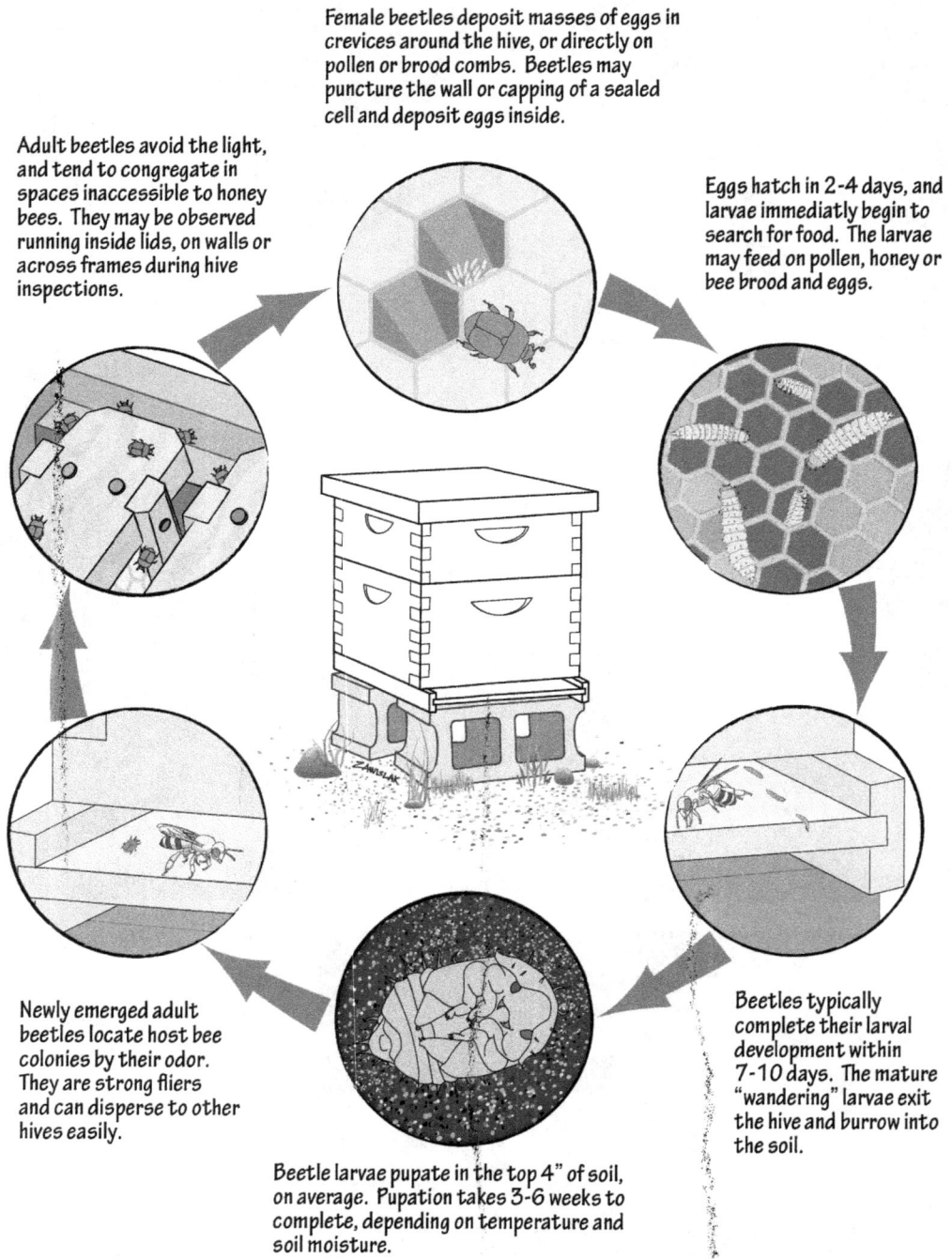

Female beetles deposit masses of eggs in crevices around the hive, or directly on pollen or brood combs. Beetles may puncture the wall or capping of a sealed cell and deposit eggs inside.

Adult beetles avoid the light, and tend to congregate in spaces inaccessible to honey bees. They may be observed running inside lids, on walls or across frames during hive inspections.

Eggs hatch in 2-4 days, and larvae immediatly begin to search for food. The larvae may feed on pollen, honey or bee brood and eggs.

Newly emerged adult beetles locate host bee colonies by their odor. They are strong fliers and can disperse to other hives easily.

Beetles typically complete their larval development within 7-10 days. The mature "wandering" larvae exit the hive and burrow into the soil.

Beetle larvae pupate in the top 4" of soil, on average. Pupation takes 3-6 weeks to complete, depending on temperature and soil moisture.

Fig. 12. Life Cycle of the Small Hive Beetle. Source: Jon Zawislak, University of Arkansas Division of Agriculture, Little Rock, Arkansas, USA.

crevices around the periphery of the inside of a highly populated bee colony, but they will lay eggs in the brood area or on pollen, if unhindered by adult bees. Most beetle eggs hatch in about 3 days but incubation period can continue for up to 6 days. Egg hatching viability is affected by relative humidity.

Beetle larvae are creamy-white in color and emerge from the egg through longitudinal slits made at the anterior end of the egg. The larval period lasts an average 13.3 days inside the bee colony and 3 more days in the soil. One US bee scientist reported beetle larvae completing maturity in 5-6 days under favorable conditions (Eischen et al. 1999). Beetle larvae are about 1 cm (0.4 inch) in length when fully grown. The length of mature larvae is variable with smaller larvae maturing slower and reaching less length on poorer diets. Beetle larvae have characteristic rows of spines on the back and have three pair of small prolegs near the head which distinguishes them from greater wax moth larvae. Another distinguishing characteristic of the two pests is that wax moth larvae leave behind a webbing mass and webbing is absent when only beetle larvae are present. Small hive beetle larvae leave behind a slimy appearance on comb. Wax moth larvae and small hive beetle larvae are found often in the same colony. The beetle larval stage is by far the most damaging life stage of the beetle (Hood 2015).

Both beetle larvae and adults prefer to eat bee eggs and brood but they also eat pollen and honey. Although small hive beetle larvae are considered the most destructive stage, the presence of adult beetles has been reported to reduce flight activity in western honey bees. (Ellis et al. 2003b; Ellis, A. and Delaplane 2008).

Mature larvae exit the hive in late evening from 1900-2200 hrs with peak activity at 2100 hrs. In the honey house, the relative humidity plays a key role in beetle larval development, so the manipulation of the moisture level can be easily integrated into an effective small hive beetle management program.

After exiting the colony, mature small hive beetle larvae enter the soil normally within about 0.6 m (2 feet) radius of the colony entrance to pupate. Pettis and Shiminuki (2000) reported that most larvae entered the soil to pupate within 1 m (39.4 inches) of the hive and to a depth of 10-20 cm (4-8 inches) in experiments conducted in sandy and loose soil in Florida, US. However, small hive beetle larvae are very mobile and could traverse a greater distance from the colony if necessary to find suitable soil to pupate. Buchholz et al.

Fig. 13. Greater wax moth larva on top and small hive beetle larva on bottom Source: Keith Delaplane, University of Georgia.

Fig. 14. Small hive beetle larva showing three pair of prolegs below and spines on top. Source: Pest and Diseases Image Bugwood.org library, Bugwood.org

Fig. 15. Small hive beetle larvae on slimed honeycomb. Source: Jamie D. Ellis, University of Florida, Gainesville, Florida, USA

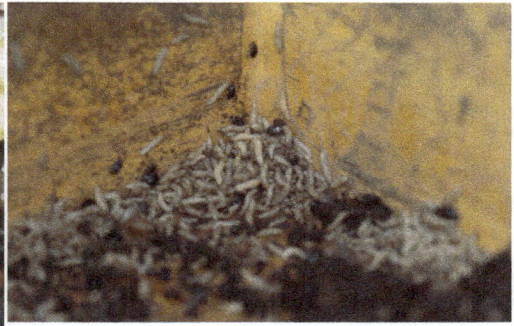

Fig. 16 (left). Small hive beetle larvae in cells. Source: Jeffrey W. Lotz, Florida Department of Agriculture & Consumer Services. Bugwood.org

Fig. 17 (above). Mature small hive beetle larvae in corner of brood box just prior to exiting hive entrance to pupate in the soil. Hood photo.

(2009) recorded wandering larvae to survive for 35 days on unsuitable soil for pupation. This high longevity of beetle larvae seeking a suitable pupation site is in line with observations that larvae can survive for 48 days without soil (Cuthbertson et al. 2008).

Laboratory tests using more dense clay soil found wandering larvae to travel and attempt to find less compact soil to pupate (Ellis et al. 2004; Meikle and Diaz 2012). Sanford (1998) reported wandering larvae have the ability to migrate long distances up to 200 m (220 yards) to find suitable soil to pupate. De Guzman et al. (2009) reported that small hive beetles successfully developed in various soils in field tests conducted in Mississippi and found that vertical movement in soil was not influenced by soil type.

The small hive beetle pupal stage lasts about 8 days or longer depending on environmental conditions. Beetle pupae were found to remain in the soil for up to 78 days (Stedman 2006; Bernier et al. 2014). Female beetles pupate slightly faster than males. Young pupae are white to brown in color and are

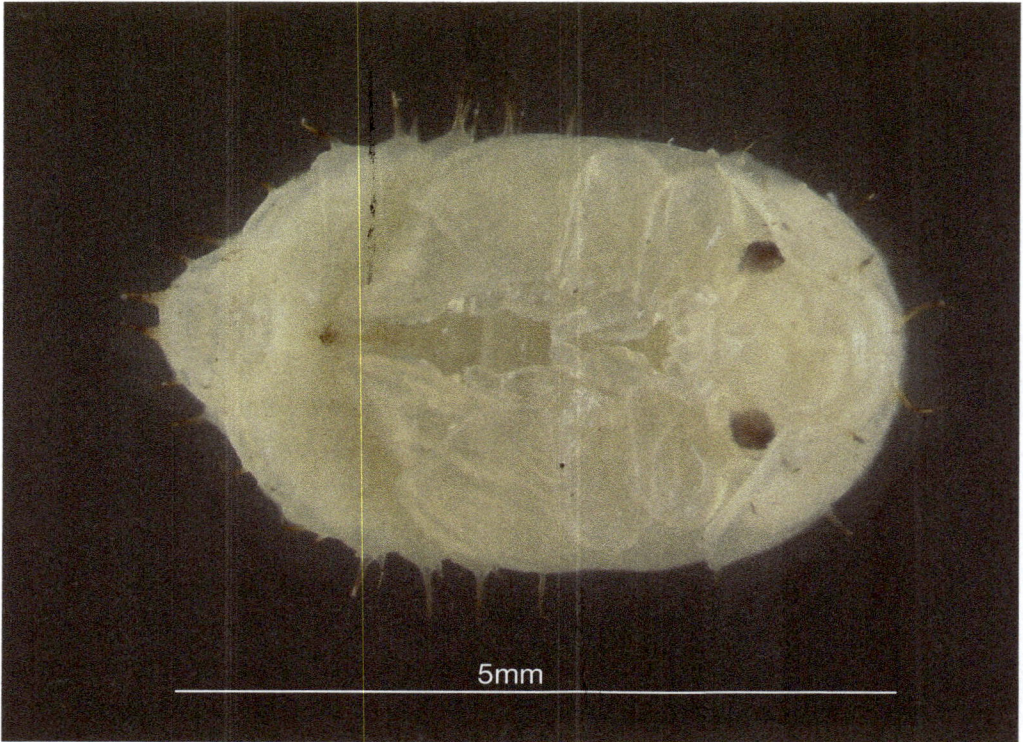

Fig. 18. Small hive beetle pupa. Source: Jamie Ellis, University of Florida, Gainesville, Florida, USA.

mostly affected by soil moisture rather than soil type. Soil type was found to have little effect on pupation survivability (Ellis 2004). Dryer soils impede pupation success rates. Pupation rates ranged from 92-98% in various soil types provided the soil was moist. This implies that beetle pest problems can be expected regardless of soil type in areas where soil moisture remains high. Therefore, soil moisture appears to be a major limiting factor in beetle regeneration thus population buildup. This may explain partly why small hive beetles are not a major problem in honey bee colonies in sub-Saharan Africa because much of Africa (except equatorial Africa) is semi-arid to arid. The dryer soil conditions would be expected to have a negative effect on beetle pupation rates. Soil density was found to affect pupation rates also with high density soils having a negative effect on pupation rates. The duration and success of small hive beetle pupation are affected by temperature, soil texture and humidity (Lundie 1940; Schmolke 1974, Ellis et al. 2004; Haque and

Fig. 19. Small hive beetle mature larvae, pupae, and young adults (brown) in ground. Source: Amanda Frake, USDA ARS, Baton Rouge, Louisiana, USA.

Levot 2005; de Guzman and Frake 2007; Meikle and Diaz 2012). Ellis et al. (2004) reported only 8.5% mortality with 23 days required to complete pupation in sandy moist soil and warm temperatures (24.6±1.3° C; 76.3±4.1° F). Meikle and Patt (2011) and Bernier et al. (2014) estimated a minimal pupation developmental temperature to be between 10 and 13.2°C (50° and 55.8°F), depending on soil water content.

Six generations of small hive beetles per year are possible under moderate US and South African climatic conditions, but this could be higher in tropical climates (Neumann et al. 2016b). De Guzman and Frake (2007) reported that almost 16 small hive beetle generations are possible a year when temperatures remain constant at 34 °C (93.2°F).

Following emergence from the soil, newly emerged adult beetles seem to prefer long-range flights, because they appear to ignore neighboring poten-

Fig. 20. Frame from dead bee colony due to starvation and low temperatures in upstate South Carolina. (See photo below of same frame) Hood photo.

Fig. 21. Closeup view of dead bees with dead adult small hive beetles in center cells. Hood photo.

tial host colonies of both *Apis mellifera* and *Bombus impatiens* (Neumann et al. 2012). Massive numbers of small hive beetle adults perhaps produced from feral bee colonies have been known to invade and disrupt apparently healthy colonies. Small hive beetles are known to aggregate especially on bottom boards (Spiewok et al. 2007). Beetle longevity and ability to mass reproduce on food materials found inside honey bee colonies have been investigated. A single mated female beetle reared on a diet of only pollen produced 591 larvae (Ellis et al. 2002a). Beetle adults survived 180-188 days when fed honey and pollen but only 19 days when fed water and beeswax. Adults feeding on honey have been reported to survive 176 days but are not likely to reproduce. Various results were reported from studies on small hive beetle longevity when beetles were deprived of food and water; results of 2 days and 10 days were reported; 7 days longevity were reported when adults emerged from the soil and were deprived of food and water. This suggests that beetles newly emerged from the soil may live for several days in search of a new host bee colony or other food source. Small hive beetles normally overwinter only in the adult stage in temperate regions and are found within the center of the honey bee colony cluster where they find food and warmth. If an overwintering bee colony dies from starvation, the adult beetles die too from cold temperatures because they are unable to fly and invade nearby live colonies in winter (Hood 2011).

Research conducted in the US reported differences in mortality rates in honey bee colonies of different European origin in October during peak small hive beetle infestations (de Guzman et al. 2006, 2010). Higher colony mortality rates were reported in colonies of Italian origin when compared to colonies of Russian origin. Further studies of Italian and Russian bees indicated no differences with respect to detection and removal of brood infested with small hive beetle eggs and larvae (de Guzman et al. 2008).

Honey bee colonies may abscond (non-reproductive swarming) as a response to heavy small hive beetle invasions, especially in African bee colonies (Hepburn and Radloff 1998). Although, it has been reported that strong African bee colonies can tolerate large beetle infestations with minimum colony effects (Ellis et al. 2003a). However, it appears there must be an upper threshold of the number of beetles necessary in a colony that will trigger the absconding event (Neumann et al. 2016b). Absconding as a result of beetle invasion has

Fig. 22. Small Hive Beetle Adults. Source: USDA (ARS.USDA.gov.)

also been reported in European honey bee colonies (Ellis et al. 2003a; Villa 2004) but likely at a lesser rate because African subspecies are more prone to abscond than European colonies (Hepburn and Radloff 1998). The increased efficiency in preparation for absconding by African bees in response to beetle invasion likely contributes to their being more resistant/tolerate of small hive beetles (Spiewok et al. 2006b).

Seasonality of small hive beetle reproduction varies from one region to another. A peak in the mean number of adult beetles captured in Hood beetle traps at Clemson University, Clemson, South Carolina, Southeastern USA, occurred in July and August (Hood 2006; Nolan and Hood 2008; Nolan and Hood 2010; Peterson 2012). A peak in the number of captured beetles in baited traps near La Cross, Florida, USA occurred during the months of May and June (Arbogast et al. 2010). Berry (2009) reported greater numbers of beetles occurred during September and October in Athens, Georgia, USA. Small hive beetle population varied throughout the year in Gabriel, Louisiana, USA, but peak numbers of beetles were observed during September and October (de Guzman et al. 2010). In the Louisiana investigations, small hive beetle populations were significantly correlated with the number of hot days, but not humid days or percentage of rainy days. In general, it appears that in the southeastern US, small hive beetle populations begin to increase in spring and build through summer and peak in late summer or early fall, then decline in late fall and winter. Strauss et al. (2013) reported a variation in peak beetle population in Gauteng, Africa with considerably higher numbers in winter of 2011 when compared to lower beetle numbers in winter of 2010.

On the other hand, Torto et al. (2010) reported small hive beetle trap capture rates in Kenya peaked during the rainy season with over 80% of the beetles captured during that time. Another investigation of front hive traps capturing mature beetle larvae exiting hives in Kenya over an entire season reported peak capture rates occurred during the "Kusi monsoon" season (Arbogast et al. 2012). In contrast, Lawal and Banjo (2008) reported that more beetle adults and larvae occurred inside honey bee colonies during the dry season than during the wet season in Nigeria.

In Australia, beekeepers reported small hive beetle larval damage was negligible during the months of June, July, and August (Rhodes and McCorkell 2007). However, much like the cyclical pattern reported above in the southeastern US, another Australian study found the small hive beetle population peaked in late fall and declined in winter to a minimum in spring (Annand 2011). In conclusion, there appears to be quite a variation in the seasonal population dynamics of small hive beetles depending on the region of the world.

An interesting behavior by adult small hive beetles is their trophallactic contact with adult bees, particularly young nurse bees whose responsibility is to feed bee larvae. The adult beetles are often held in captivity by their worker bee captors in various places in hives to deprive the beetles of their freedom to roam and lay eggs on brood or pollen.

The beetles are known to beg and successfully dupe their captors into feeding them droplets of food. Ellis et al. (2002b) reported that beetles on average beg for food from bees 11-12 times before they are successful in being fed and Neumann et al. (2015) reported that success in chance of being fed increases with experience. Beetles are known to survive long periods of

Fig. 23. An adult small hive beetle antennating with a honey bee or in the process of successfully receiving food from the bee. Hood photo.

time in captivity as a result of this behavior called trophallaxis. Lundie (1940) reported that beetles can survive "up to 6 months and probably longer" in sheltered host nests which provide them food and warmth.

Other odd behavior by small hive beetles inside honey bee colonies include beetles feeding on freshly emerged live honey bee workers (Pirk and Neumann 2013). Neumann et al. (2016b) reported seeing adult beetles mounting a honey bee workers' abdomen and cutting through the tissue between the tergites with its mandibles. Since small hive beetles are the only known species that are able to induce trophallactic feeding from honey bees (Ellis et al. 2002b), there is likely no competitor of adult small hive beetles in honey bee colonies (Neumann et al. 2016b).

Chapter 4
Genetic Diversity

Small hive beetle genetic diversity has been studied in beetles collected from 1996-2000 in the US and South Africa (Evans et al. 2000, Evans et al. 2003). Two distinct mitochondrial DNA (mtDNA) haplotypes were discovered from beetle collections made in southeastern US whereas 13 beetle haplotypes were described from collections made from southern Africa. Mitochondrial DNA sequence data from these separate beetle collections suggest strongly that a single species exists in both continents (Evans et al. 2000). The two beetle haplotypes found in the US (NA1 and NA2) are indistinguishable using the COI gene from native beetles collected from two countries in southern Africa: South Africa and Zimbabwe (Evans et al. 2000). While these findings suggest significant similarities between the beetles collected in the US and the southern African countries, conclusive proof does not exists inferring that the beetles found in the US could have originated from other regions of the African continent (Evans et al. 2000).

The mtDNA analysis conducted by Evans et al. (2000) and Evans et al. (2003) yielded an irregular distribution of the two small hive beetle haplotypes from collections made over the 5 year period in the southeastern US. The earliest samples collected in 1996 and 1997 in South Carolina showed haplotypes NA1. Although later samples taken from South Carolina in 1998-2000 included both haplotypes, NA1 and NA2. NA1 continued to slightly dominate. Samples collected in Georgia in 1998-2000 showed a higher frequency of NA2, and Florida samples collected over the same time period showed roughly equal haplotype frequency. Beetle samples collected from South Carolina and Georgia showed strong biases in mtDNA haplotype frequencies over the period 1998-2000 that indicates that the two haplotypes are not randomly distributed among beetle populations. Further, results from these investigations presented competing hypothesis that beetle haploid frequency differences indicate two or more separate introductions into the US, ongoing selections at different sites, or genetic bottlenecks as beetles invade new sites (Evans et al. 2003). Barring strong bottlenecks as beetles spread to new apiaries and regions in North America, regional differences in haplotype frequencies will erode reflecting the current equal frequency in Florida (Evans et al. 2003).

No correlation was found in small hive beetle size and haplotypes which indicates that no selection pressure exists between haplotype NA1 and NA2 found in the US (Evans et al. 2003). Body mass is often used as a surrogate for competitive abilities and reproductive success in insects (Messina and Slade 1999).

Chapter 5

Economic importance

The small hive beetle is considered to be of little economic importance in its native range of southern Africa where it is listed as a threat only to weakened or stressed colonies. African bees are capable of preventing or postponing the beetle from breeding in the hive as long as colonies remain strong (Neumann and Elzen 2004). Several scientists have reported that successful beetle reproduction occurred most often in weak or stressed African honey bee colonies or in recently abandoned colonies and is far less in strong colonies (Lundie 1940; Schmolke 1974; Hepburn and Radloff 1998; Neumann and Elzen 2004). In some regions of Africa, heavy small hive beetle reproduction normally appears to wait until absconding occurs (Hepburn and Radloff 1998; Neuman and Elzen 2004). In contrast, beetle infestations in the southeastern US affected even robust bee colonies which required control measures by the beekeeper. In the US and Australia, small hive beetle damage in European colonies follows a characteristic pattern (Ellis J. and Ellis, A. 2016):

1. "Adult invasion into colonies

2. Population build-up of small hive beetles

3. Reproduction of small hive beetles

4. Significant damage to brood, pollen, and honey stores by feeding small hive beetle larvae

5. Mass exodus of larvae from the colony

6. Pupation in the soil and

7. Emergence as adults and subsequent re-infestation of colonies."

Fig. 24. Small hive beetle adults alongside honey bee workers. Source: Jessica Lawrence, Eurofins Agroscience Services, Bugwood.org.

Research indicates that even strong colonies of European honey bees can be taken over and killed by small hive beetles within less than 2 weeks. Neumann et al. (2010) reported mass beetle reproduction resulting in total structural collapse of a strong bee colony in less than 10 days. Therefore, successful small hive beetle reproduction seems to be more common in strong European colonies when compared to African colonies (Neumann et al. 2010).

Small hive beetles feed on honey, pollen and brood in bee colonies and have been implicated often in both colony mortality and increased absconding rates. A quarantine on movement of honey bee colonies was established in Florida in June 1998, but it was soon withdrawn by the Florida Department of Agriculture and Consumer Services (Fore 1998).

There has been some concern that honey bee colonies may suffer from multiple stress factors making them susceptible to possible additive or synergistic effects. Research was conducted to manipulate varying levels of small hive beetles and varroa mites, *Varroa destructor*, infestations (Delaplane et al. 2010). Increasing densities of either pest when manipulated alone in colonies resulted in a predictable increase in colony demise. However, when both pests were manipulated in same colonies, varroa mite levels decreased as apiary-wide small hive beetle levels increased. These results were unexpected and indicated that small hive beetle and varroa mite infestation levels do not interact such that damage by one was affected by changing levels of the other. Other researchers conducted similar experiments and came to the same conclusion that there was a weak correlation between the combined effects of small hive beetles and varroa mites in honey bee colonies (Schafer et al. 2010a).

Small hive beetles have the potential to vector and transmit viruses from infected bees to healthy bees. Investigations have been conducted that showed that small hive beetles can transmit deformed wing virus (DWV) by becoming virus positive by feeding on DWV infected honey bees and virus contaminated foods such as honey and pollen. Test results also indicated that DWV could replicate in small hive beetles and thereby have the potential of transmitting the virus from infected beetles to healthy bees (Eyer et al. 2008).

Adult small hive beetles have also been shown to be infected with Sacbrood virus via food-borne transmission (Eyer et al. 2009)

Small hive beetles have been reported to vector the bacterial agent *Paenibacillus larvae* which is the causative agent of American foulbrood in honey bees (de Graaf et al. 2013). Both beetle adults and larvae became infected with American foulbrood spores when exposed to brood combs showing clinical symptoms (Schafer et al. 2010b). Field tests where honey bee colonies were artificially infested with contaminated adult beetles resulted in higher levels of *P. larvae* spores in adult worker bees and honey (Schafer et al. 2010b).

Small hive beetles can result in significant financial loss to beekeepers (Delaplane 1998). The estimated losses to small hive beetles experienced by beekeepers in the US in 1998 were $3 million (Elzen et al. 2001). Losses were in the form of colony destruction and damage to stored honey supers in honey houses. Some commercial beekeepers in the US reported losing thousands of bee colonies and associated equipment to beetles the first few years fol-

Fig. 25. Slimed frame top bars as a result of small hive beetle damage. Hood photo.

lowing their discovery. The small hive beetle came on the heels of major varroa mite problems in the US, especially in some large commercial operations which were reeling from the effect of the mites during the early years following the introduction of small hive beetles (Hood 2015). Van Engelsdorp et al. (2007) reported survey results that US commercial beekeepers claimed varroa mites, tracheal mites, and small hive beetles were the leading causes of their colony losses in winter of 2006 and 2007.

The small hive beetle has been devastating to the Hawaiian beekeeping industry since its first discovery there in April 2010. The economic impact caused by small hive beetles on the islands has been huge mainly due to losses in the export market of queen bees to quarantine restrictions by some countries (Robson 2012). Connor (2011) reported on a survey conducted on the Big Island in 2010 that 55% of the managed bee colonies were lost and 34% of the beekeepers who are mostly hobbyist had lost all their colonies. Eighty percent of the colony losses were attributed to small hive beetles or varroa mites which took a double whammy on the islands beekeeping industry (Connor 2011).

Fig 26. Brood frame destroyed by small hive beetles. Notice fermented honey. Hood photo.

Mixed reports are coming from Australia as to the level of damage the small hive beetle is causing in that country. Strategies for prevention and management of beetle have been developed and provided to the beekeeping industry. Initial reports indicated that beetles have not caused significant damage in Australia when compared to damage caused in the US, especially coastal southeastern US. A beetle survey conducted in managed honey bee colonies from October 2002 to January 2003 in the New South Wales area reported 120 positive detections out of more than 1,000 samples received (Gillespie et al. 2003).

More recent reports from Australia indicate that small hive beetles have been found in far North Queensland and beetles have killed about a third of individual beekeepers hives in New South Wales to the south. A survey was conducted in New South Wales that estimated losses due to small hive beetles to be about 4,500 bee colonies during the period 2002-2006 (Rhodes and McCorkell 2007). Another survey in the Queensland area indicated that more than 3,000 bee colonies had been lost to small hive beetles resulting in an estimated economic damage of approximately $1,200,000 (Mulherin 2009).

Seven drought years in the Australian beetle-infested areas restricted movement of colonies which may have resulted in slowing the spread of the beetle initially, but since then the beetle appears to be showing up in new areas and causing greater damage. The major losses have been a result of the negative effect on overseas and domestic package and queen bee markets. One report from Australia claimed that stressed bee colonies suffering from European foulbrood are prone to result in major small hive beetle problems (White 2003). Overall, the small hive beetle has proven to be a serious economic threat to the Australian beekeeping industry given suitable conditions, especially in eastern Australia (Annand 2008; Spooner-Hart et al. 2016).

The US queen and package bee production industry has been challenged at a higher level from small hive beetles given that their operations are mostly stationary and located in warmer regions of the country, which adds to their vulnerability to this pest (Hood 2015). Beekeepers also have concern over receiving small hive beetles when purchasing queens and package bees, especially from suppliers located in beetle infested regions.

Concern over spread of small hive beetles to the UK has been reported. Favorable conditions required for beetle survival are met in many areas of the UK. Therefore, the risk management recommendations for small hive beetles in the UK include prohibition of bee imports from infested countries (Brown et al. 2002).

In Canada, small hive beetles have not become well established except in parts of Ontario, mainly as a result of the countries' unfavorable climatic and environmental conditions. However, the commercial impact on trade relations and movement restrictions has been significant for many Canadian beekeepers and their industry (Neumann et al. 2016b).

The Calabria area of southern Italy has been negatively impacted by the invasion of small hive beetles into that region. The commercial beekeepers in the region can no longer ship queens and package bees to beekeepers outside the beetle-infested region or to other European countries. The fruit pollination business in the region has been served a major setback as a result of reluctance on beekeepers ability move bee colonies freely in and out of the region. There have been great losses to beekeepers by having their bee colonies destroyed in the beetle eradication efforts by the Italian authorities (Quigley 2015a, 2015b).

Concern over possible small hive beetle damage to other commodities such as fruit has been raised both in the presence and absence of honey bee colonies in the field (Buchholz et al. 2008). Scientists have investigated beetle reproduction on alternate food sources. Adult beetles fed on honey lived for over 5 months whereas those feeding on fruit lived for over 2 months (Ellis et al. 2002a). Beetles regenerated when offered a diet of avocado, cantaloupe or grapefruit in confinement. Laboratory reared beetle adults were fed rotten and fresh kei apples, *Dovyalis caffra*, and survived an average 58.6 days and 63.9 days, respectively (Ellis et al. 2002a). Average number of offspring produced from three mating beetle pairs after feeding on rotten Kei apples in laboratory tests were significantly less than the average number of offspring produced from three mating beetle pairs feeding on pollen comb (10.6 vs.1,096.4) (Ellis et al. 2002a). Successful beetle reproduction in laboratory experiments has been reported on other fruits such as banana, pineapple and mango (Keller 2002); grapes (Buchholz et al. 2008); green grapes and oranges (Arbogast et al. 2010); and on decaying meat (beef schnitzel, Buchholz et al. 2008). The number of offspring per breeding couple was lower on these fruits compared to the number of offspring produced on hive products (Keller 2002; Buchholz et al. 2008).

Buchholz et al. (2008) conducted laboratory research to investigate whether small hive beetles would use alternative food sources (fruit and meat) when hive products from honey bee colonies were also available. A combination of bee brood, honey and pollen was always the preferred food in choice tests for egg oviposition and larval feeding. However, when only honey and pollen (without bee brood) were presented, bananas were chosen most often, which was the case for experiments on larval food preference and adult oviposition choice (Buchholz et al. 2008).

The poor reproductive success of beetles feeding on fruits is likely a result of minimum nutritional requirements being met, but there is the possibility of beetle regeneration on fruit in the wild when no bee colonies are present. Although, no record exists which reports successful beetle regeneration on fruits or vegetables in field conditions. Current data evidently show small hive beetles in principle can exploit an abundance of alternative food resources, but there is extremely little evidence that they do so in the field (Neumann et al. 2016b). Since small hive beetles can survive for several days on various fruits, there exists a strong possibility that beetles can be transported by fruit truck or cargo shipments to non-infested regions of the world. In summary, there are risks involved in small hive beetle movement pathways associated with various fruit and these cannot be completely discounted, however a preponderance of evidence exists that they pose a much lower risk when compared to the movement of bees, beekeeping equipment and bee related products.

Small hive beetles can be problematic for honey bees kept inside observation hives (glass enclosed hives) which are used for school demonstrations, county fairs, festivals and other events to showcase honey bees. Observation hives are prone to swarm more often than larger hives simply because of smaller quarters for brood rearing and food stores. Swarming usually contributes toward stressful colonies and observation hives also provide additional cracks and other hiding places for beetles. More routine replacement of colonies in observation hives results in extra expense and labor for beekeepers.

Fig. 27. Observation Hive at Clemson University. Bellinger Photo.

Bumble bees and other non-*Apis* species are additional concerns as possible threat to small hive beetle invasion and may serve as alternative hosts (Neumann et al. 2008; Hoffman et al. 2008). Bumble bees are native to the Americas, Europe and Asia (Michener 2000) but are not found in sub-Saharan Africa. Therefore, the possible host/pest interactions between bumble bees and small hive beetles has been rather short-lived, dating back to about 1996 when beetles were first collected in North America (Hood 2011).

In controlled studies, small hive beetles regenerated on colonies of bumble bees, *Bombus impatiens* (Spiewok and Neumann 2006a). Artificially beetle infested colonies of *B. impatiens* produced fewer live bees, yielded more dead adult bees and increased comb damage more than control colonies (Neumann and Elzen 2004). Adult small hive beetles were reported to naturally infest commercial *B. impatiens* colonies in the field (Spiewok and Neumann 2006a) and in greenhouses (Hoffmann et al. 2008). Small hive beetles were found to preferentially oviposit next to brood pots. *B. impatiens* defended their colonies against small hive beetles by removing and attacking both adults and larvae (Hoffmann et al. 2008).

These investigations were conducted in non-natural conditions with bumble bee nests placed above ground as opposed to wild bumble bees that nest in the ground. No one has reported finding small hive beetles in natural bumble bee colonies, but surveys have not been conducted to refute this possibility. When beekeepers move beetle-infested honey bee colonies from location to location for commercial pollination purposes, they may leave behind great quantities of beetle pupae in the soil which emerge to seek and find a suitable food source, such as a nearby managed honey bee colony or a nearby feral honey bee colony to invade and overwinter. There is, however, a possibility that small hive beetles may be attracted to ground nesting bumble bees because of similar odors (bee brood and honey) as honey bee colonies. This could prove detrimental to bumble bee colonies during the warmer seasons of the year. However, bumble bees do not overwinter as colonies in many regions of world, so the beetles presumably would perish along with the colonies for lack of food and warmth in winter. Mated queen bumble bees are the only overwintering life stage that survive in protected quarters in temperate regions of the world. The queens have to re-establish a new colonies in the spring of the year. In summary, field data are too scarce to draw any definitive determination about the potential role of bumble bees as alternative hosts of small hive beetles (Neumann et al. 2016b).

Stingless bees which thrive in many warmer regions of the world are another threat to small hive beetle invasion. In West Africa, natural infestations of small hive beetles have been found in colonies of stingless bees, *Dactylurina staudingerii* (Mutsaers 2006; Williams 2015). Pena et al. (2014) reported that managed and wild colonies of *Melipona beecheii* had been infested with small hive beetles in Cuba with seven out of 258 surveyed hives had adult beetles and two hives had both beetle adults and larvae. Beetle infestations were also found to be associated with damage to recently founded *M. beecheii* colonies (Pena et al. 2014; Neumann et al. 2016b). Studies have been conducted that report small hive beetles can naturally infest stingless bee colonies under certain circumstances such as disturbed or newly founded colonies. Stingless bees have some unique protective measures to guard against intruders such as small hive beetles. Rapid live mummification was reported to effectively prevent beetle advances and remove their ability to reproduce by stingless bees, *Tetragonula carbonaria* (Greco et al. 2010). Similar to experiments with *T. carbonaria*, non-invasive tests with colonies of *Austroplebeia australis* resulted in beetles being entombed alive by worker bees within 6 hours (Halcroft et al. 2011). Undisturbed strong colonies of stingless bees apparently have the necessary defense mechanisms to protect themselves from beetle invasion (Halcroft et al. 2011; Greco et al. 2010). As with bumble bees, additional research is needed with stingless bees to determine the seriousness of this new colony invader.

In conclusion, small hive beetles have contributed to great losses to the beekeeping industry as a result of time and labor to detect and control the beetles in many regions of the world (Calderon et al. 2006). Losses in managed honey bee colonies and associated equipment, honey production and pollination are the main economic casualties suffered by the beekeeping industry. Delaplane and Mayer (2000) noted a recent decline in bee numbers can be attributed to various bee pests and diseases, including the small hive beetle. The bee decline will no doubt result in a significant adverse effect on pollination habitats where plants rely on bees. A potential decline in the native bee fauna including honey bees, bumble bees, and stingless bees as a result of small hive beetles will have a negative impact on bee biodiversity (Cuthbertson and Brown 2009).

Chapter 6

Small hive beetle control

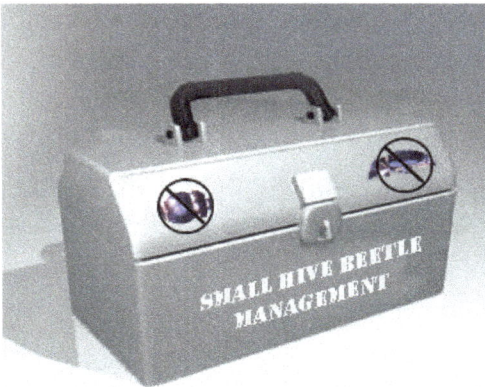

There are currently many options available to beekeepers to manage small hive beetles. In the remainder of this book we are going to discuss the integrated pest management (IPM) approach to small hive beetle control. First, a broad agricultural definition of "integrated pest management" is given by Knipling, E. F. (1979) who earlier referred to this system of control as "integrated pest control" or more broadly "insect pest management" as insect control "based primarily on close monitoring of insect conditions on various crops and the use of control measures where and when necessary with emphasis on methods that will permit natural control agents to have their maximum effect in regulating the population. When control measures are necessary, however, this generally involves the use of chemical insecticides. The system emphasizes the establishment of economic threshold levels for many species on various crops as a basis for deciding when the application of insecticides is warranted." Nasr, M.E. (2001) gave a more current and brief definition of integrated pest management: "the use of compatible combinations of genetic, cultural, regulatory, physical, and chemical methods to manage pests in an economically and environmentally sound manner."

Knipling (1979) mentioned the use of an economic threshold which can be defined as "the infestation level at which control measures should be implemented to prevent an increasing pest population from reaching the economic injury level" (Nasr 2001). The economic injury level is "the lowest pest level that will result in economic damage and justify the cost of control" (Nasr 2001).

We are somewhat hampered in our integrated pest management approach to controlling small hive beetles, given that we do not have economic thresholds or economic injury levels to guide our management decisions. However, we do have many control options (prevention, genetic, mechanical, physical, biological, chemical, etc.) available that allow us to manage small hive beetles in a responsible manner without contaminating honey and other hive products or endangering the beekeeper.

Although we do not know all the answers to what conditions favor small hive beetle reproduction, scientists and beekeepers have developed an arsenal of control recommendations and tools through research based on the beetles behavior and biology (Hood 2015). We are going to discuss the integrated management of small hive beetles in the context of the eight basic "IPM Beekeeping Principles" (Hood 2009a) that include: preventive practices, acceptable pest levels, monitoring practices, genetic control, mechanical control, physical control, biological control, and chemical control.

Efficient control of small hive beetles rarely depends on a single method of control, but may require the use of several control measures. Beekeepers should become familiar with all available control measures and select the ones which give the most efficient beetle control, but at the same time result in minimum environmental impact. A listing and discussion of many options that are available to beekeepers currently to control small hive beetles are found below.

Preventive Measures to Control Invasive Pathways

We will discuss first the national and international measures that are recommended to prevent small hive beetle invasion into new regions of the world. The first known small hive beetle invasion outside of its native sub-Saharan Africa into the coastal southeastern US in 1996 occurred mostly unnoticed for about 2 years. This gave the beetle time to gain a stronghold in its new territory of favorable climate and environmental conditions. Most of the beekeeping world other than beekeepers and authorities in South Africa had never seen nor heard of this soon to become serious international honey bee pest during the next 20 years. Following this 20 or so year invasive period, we now know much more of the behavior and biology of this pest as well as recommendations on how to control this hive pest. However, there is much more to learn and many questions go unanswered, such as how to prevent small hive beetle entry into new regions of the world.

The World Animal Health Organization (OIE) report published in 2012 made several recommendations for minimizing the risk of introducing small hive beetles that are associated with bees, apicultural equipment, and hive products. The list of recommendations included inspection of consignments, transporting live bees only from areas known to be free of *Aethina tumida*, covering bee consignments with fine mesh to keep beetles out, cleaning of equipment and freezing honey (OIE 2012).

Knowledge about the likelihood of different invasive pathways is essential to prohibit the further spread of small hive beetles to new areas of the world through adequate legislation and border control measures (Neumann et al. 2016b). Major seaport entry has been a likely introduction route of small hive beetles into some countries: USA (Hood 2000, 2004), Italy (Mutinelli et al. 2014), Australia (Ellis and Ellis 2016). However, no conclusive evidence exists on how or when small hive beetles may have been introduced in either of the three introductions listed above.

Possible modes of small hive beetle entry into new regions of the world include queen bee cages, package bees, honey bees on comb in managed colonies whether full size or 5-frame nuclei, hitch-hiking swarms, beeswax, slumgum, other beekeeping supplies and equipment, and soil or compost associated with the plant industry trade. Colonies of *Bombus* spp. and stingless bees provide another possible source of beetle entry.

Although small hive beetles are not thought to be present in the UK, the National Bee Unit (NBU) and its inspectors have increased surveillance programs in hopes of intercepting possible exotic pests including the small hive beetle (Williams 2015). They are well aware of the need for early detection of small hive beetles for any chance of eradication. "At risks" locations in England and Wales include the following exotic risk points: crude hive products importers, landfill sites associated with imported produce, military airports (UK forces), military airports (American), fruit and vegetable wholesale markets, freight ports/ports, freight depots, civilian airports (Williams 2015).

Long distance survival of adult and/or immature life stages of small hive beetles depend on both storage conditions and food availability during transport (Neumann et al. 2016b). Border inspections should be strongly considered for any possible beetle transport materials including suspect commodities prior to shipment and at destination sites. Examples of early detection and successful elimination of small hive beetles have included processed wax (Manitoba, Canada; Neumann and Elzen 2004) and queen cages (Portugal; Murilhas 2004). Eradication of small hive beetles will only be made possible when beetles are detected soon after their arrival and few beetles are involved. If early detection is not accomplished, adult beetles have the ability to spread quickly into the surrounding managed honey bee colonies as well as the wild or feral colonies making eradication all but impossible (Hood 2015).

Migratory beekeeping or movement of managed colonies has played a major role in further spread of small hive beetles once introduced into a country, especially when detection has been delayed: USA (Hood 2000; Zawislak 2014; Wikipedia 2017), Quebec, Canada (Evans et al. 2003; Giovenazzo and Boucher 2010), Coahuila, Mexico (Del Valle Molino 2007) and Perth, Australia (Neumann and Elzen 2004).

Several molecular techniques have now been used to diagnose small hive beetles (Evans et al. 2000, 2008; Lounsberry et al. 2010; Ward et al. 2007). Ward et al. (2007) used real time PCR in conjunction with an automated DNA extraction protocol to screen hive debris for the presence of small hive beetles. The same protocol was effective in detecting DNA from small hive beetle eggs, larvae and adult specimens which were collected in Africa, Australia and North America and to detect beetle DNA extracted from spiked and naturally beetle infested debris.

These modern genetic tools can now be used to trace back the origin of invasive species, which can be used to mitigate future introductions (Neumann et al. 2016b). Evans et al. (2000, 2003) reported the use of small hive beetle mt-DNA sequence analysis to determine that populations of beetles collected from the US and South Africa are of the same species. However, in their investigations it was unclear whether single or multiple introductions were made into the southeastern US. Further studies showed that beetles found in Australia have a different origin than beetles found in the US (Evans et al. 2000, 2003; Lounsbury et al. 2010). The small hive beetles detected in Quebec, Canada appear to have originated from the US (Evans et al. 2003, 2008) whereas small hive beetles discovered in package bees in Alberta, Canada in 2006 were confirmed to have originated from Australia (Lounsberry et al. 2010; Neumann et al. 2016b).

National and international beekeeping trade no doubt has played a major role in spread of this invasive pest (Neumann et al. 2016b; Alberta, Canada, Australian package bees, Lounsberry et al. 2010). From an economic point of view, it seems prudent therefore to focus on strengthening legislation and to promote stricter control of international bee trade to curb the future global spread of small hive beetles (Neumann et al. 2016b). On the other hand, it will also be wise to closely monitor the international movement of those commodities that are known to sustain small hive beetles such as fruit and soil in plant trade.

Preventive Cultural Practices for the Beekeeper

Beekeepers are advised to maintain strong, healthy colonies in areas where small hive beetles are found. Beekeepers should practice good colony management to help the bees defend their colony from the negative effects of hive pests such as the small hive beetle (Westervelt 2005; Hood 2011). Weak honey bee colonies are more vulnerable to small hive beetles simply because they lack enough bees to protect their brood and pollen which are needed resources for beetle reproduction (Hood 2015). Good colony management starts with a good laying queen that can regulate the colony population to maximize their chances of survival. Her genetic makeup is paramount in that her progeny must be able to sustain the colony in the presence of various diseases and pests, including small hive beetles. *In general, a high bee-to-comb ratio is recommended for small hive beetle control.* Fletcher and Cook (2005) noted that hives where the bees do not exhibit good hygienic behavior (attempt to remove from the hive beetle adults and larvae) toward beetles should be re-queened or replaced by bees that do.

Fig. 28. A healthy queen is an integral part of a strong honey bee colony. Source: Bellinger Photo.

Fig. 29. A strong, well populated honey bee colony is recommended for small hive beetle control (high bee-to-comb ratio). Hood photos.

Fig. 30. Adult small hive beetles are often seen running across the combs in an effort to evade honey bees. Source: J. Ellis, University of Florida, USA.

The beetle larval stage is by far the most harmful life stage to a honey bee colony, so efforts should be made to focus on limiting reproduction of beetles in the colony (Neumann et al. 2016b). When strong colonies are infested, the adult beetles are conducting a "sit and wait" reproductive strategy as they are many times confined by guard bees or are hiding in gaps or other places in the hive where needed protein for larval development is rare or absent (Neumann et al. 2016b). Only if the hive situation changes which results in the adult beetle freedom from captivity allowing them to roam freely in the hive will they try to lay eggs close to a larval food source (bee brood or pollen). Beekeeper over-manipulation of a marginally strong hive may contribute to this situation along with poor frame re-placement or poor frame alignment in the hive.

Another cultural technique recommended for beetle control is the placement of colonies in full sun to create drier soil conditions to help prevent successful beetle pupation in the ground. Beetles need moist soil to successfully pupate and the placement of colonies in shady, damp locations is not recommended. This recommendation runs counter to what most beekeepers were taught in the past: to place colonies in locations that offer early morning sun and afternoon shade, particularly in the hot summer months. Beekeepers should also be careful in placement of their colonies in or near irrigated crops which are often grown in damp soil conditions.

Acceptable Pest Levels

Although attempts have been made to develop a treatment threshold for small hive beetles in managed colonies, there has yet to be one published. Research is also needed to develop an effective beetle sampling tool which

Fig. 31. These two sites would not be good apiary locations, if beetles are present. Hood photo.

will accurately estimate the total number of beetles in a colony without having to conduct a whole colony beetle count. We are somewhat handicapped in our IPM approach to control this hive pest without a treatment threshold system. However, there are some general guidelines that are recommended to manage this hive pest.

The beekeeper must resist the temptation of treating the colony with a pesticide when only a few beetles are present in the hive or treating when it is obvious the colony collapse level has been reached. At times, it is best to allow the honey bees to have the freedom to protect the colony from pests without beekeeper intervention. Honey bees have their own methods of defending the colony from the harmful effects of small hive beetles.

Three lines of defense are proposed as a "tug of war" of interaction between the host honey bees and the invading small hive beetles (Neumann et al. 2016b). The first line of defense is the guard bees at the hive entrance. Fewer adult small hive beetles were reported inside colonies having smaller modified hive entrances (Ellis et al. 2002d). This suggests that guard bees are more successful in preventing adult beetles from entering the colony when having to protect a smaller entrance. African bees are known for their increased aggressiveness toward beetles to protect the colony compared to European bees (Neumann and Elzen 2004), but Hepburn and Radloff (1998) noted that reduced chance of colony invasion by this subspecies is likely a result of the use of propolis to reduce the size of the colony entrance.

The second line of defense is the host worker bees that patrol the nest and guard combs (Neumann et al. 2016b). This patrolling behavior is more pronounced in the brood area of a colony and less pronounced in the outer frames and honey supers (Schmolke 1974; Solbrig 2001). Spiewok et al. (2007) reported small hive beetle distribution is influenced by presence or absence of bees within the brood nest area.

The third line of defense of a bee colony is the removal of beetle eggs and larvae by host worker bees (Newmann et al. 2016b). This line of defense is more pronounced in African bees than European. When a colony fails to patrol beetles and prevent them from entering the brood nest, beetles respond by laying eggs on the brood or in gaps to protect their offspring (Neumann and Hartel 2004). Mass oviposition by beetles in the unprotected brood can lead to total colony collapse (Hepburn and Radloff 1998). To prevent this massive beetle reproduction in the brood area, the adult bees can better patrol and prevent egg laying from occurring or the bees can detect and remove eggs once they are laid. One means of prevention of egg laying in the brood is called social encapsulation of "corralled" small hive beetles (Neumann et al. 2001) which may be in part due to the amount of propolis utilized in the hive. African bees are reported to use more propolis in the hive than European bees which likely contributes to an increased number of confinements per colony and an increase in number of beetles encapsulated in African colonies (Neumann et al. 2001). Both scientists and bee breeders are optimistic that any bees displaying defensive traits toward small hive beetles and/or the ability to incarcerate beetles in propolis prisons, will be selected for in the future to aid in development of resistance (Neumann et al. 2016b).

In summary, honey bees have their own methods of controlling pests and disease. Sometimes it is best to allow this natural process to work without beekeeper intervention, especially when low level beetle numbers are present. However, beekeepers are warned to closely monitor their colonies when conditions are favorable (warm and rainy weather) for beetle reproduction and to intervene in an appropriate manner when necessary.

Minimum Manipulation

As mentioned above, worker bees chase and corral adult beetles into confined areas inside the beehive which prevents the beetles from freely roaming the hive and laying eggs on or near stored pollen and bee brood. The beetles need the pollen and brood as a source of protein for sustained nourishment and growth. Without the necessary protein in their diet, beetle reproduction is hampered. Both scientists and beekeepers have reported the observation that an increase in beetle larval damage occurs often following colony inspections.

When beekeepers open their colonies, beetles often escape confinement and freely roam the colony. If the colony is showing signs of stress, the bees may not be able to re-corral the beetles, which may lead to an increase in beetle reproduction. Beekeepers should not open their colonies unnecessarily. This is particularly true during times of the year when beetle populations tend to increase which begins as early as May in the southeastern US and may continue till early fall. New beekeepers should resist the temptation to over manipulate their colonies. The queen status simply does not have to be checked on a daily basis. Leaving beehives open too long during colony inspections can also lead to stress from robber bees from nearby colonies, especially during times of dearth.

Fig. 32. Correct frame alignment is recommended when replacing frames in a beehive for small hive beetle control. Source: R. Bellinger.

Over manipulation of bee colonies has often been reported to lead to additional beetle problems perhaps due to poor frame replacement within the brood chamber, especially by novice beekeepers. Frames placed too close together which leads to a lack of worker bee patrolling may result in unhindered beetle activity. Beetles can freely oviposit on the comb in these congested areas and escape pursuit by worker bees (Neumann et al. 2016b). Therefore, beekeepers should be extremely careful when replacing brood frames to insure correct frame placement.

According to Annand (2011), small hive beetles can seriously impact honey bee colonies that are compromised, such as diseased, drone layers or subject to poor beekeeping management practices. This again supports the idea that poor colony management by beekeepers can be "a trigger for small hive beetle reproductive success" (Neumann et al. 2016b.)

In beekeeping operations that have a history of beetle problems, it is recommended not to use hive inner covers (crown boards) or end frame spacers as they provide additional hiding places for the beetles to hide and avoid bee contact and imprisonment.

Monitoring Practices on a Large Scale

The use of small hive beetle monitoring tools are recommended at high risk areas that are subject to importation of apicultural products and commodities that sustain small hive beetles, such as ports or airports. Early detection of small hive beetles is of paramount importance to have a high chance of pest eradication. This is particularly true in those areas of the world which have conditions favorable for beetle reproduction such as the Caribbean, South and Central America, and Southeast Asia.

Sentinel honey bee colonies are recommended in high risk zones where small hive beetles may enter new regions. Careful routine inspections of sentinel colonies by trained personnel is very important to provide early detection (Chauzat et al. 2015). This procedure can be time consuming but it is also highly accurate (Neumann et al. 2016b).

Molecular techniques have been described to diagnose small hive beetles (Evans et al. 2000, 2008; Ward et al. 2007; Lounsberry et al. 2010). Ward et al. (2007) used real-time PCR in conjunction with an automated DNA extraction protocol to screen hive debris for presence of small hive beetles (Neumann et al. 2006b). This method detected DNA from beetle eggs, larvae and adult specimens from Africa, Australia and North America and to detect beetle DNA extracted from spiked and naturally infested debris (Neumann et al. 2006b). A surveillance study in Spain used this method of screening hive debris to evaluate the possible presence of small hive beetles in the country (Cepero et al. 2014).

Monitoring Practices for the Beekeeper

If small hive beetles are present in a colony, their presence is normally obvious when the beekeeper removes the hive top and carefully inspects underneath the top and exposed frame top bars during warm weather. Beetles do not care for light conditions and will seek a dark refuge quickly. So, the beekeeper can often get a good idea of the number of beetles present in the colony simply by checking for beetles in the top of a hive. If there are many beetles in the top of a hive, a further inspection of the brood chamber is highly recommended to get a better idea of the total beetle population, especially the presence of beetle larvae.

Another quick beetle monitoring population tool is to lift the top super off the colony and bounce it gently a couple of times on an overturned telescoping hive top which the beekeeper has placed on the ground. If beetles are present in the super, some will dislodge and fall to the hive top inner surface below. Another similar sampling method (see photo below) is to allow the honey super to remain in place on the overturned telescoping hive top for approximately 10 minutes (Zawislak 2014). Adult beetles are light sensitive and will be left behind on the overturned hive top. Beekeepers are cautioned to practice with care the latter method because it may lead to robbing from nearby stronger colonies, particularly during periods of dearth because colonies are more prone to rob during non-nectar flow periods.

A tell-tale sign of a major small hive beetle problem in a hive is when the entrance landing board is soiled with residues of fermented honey, pollen and beeswax which have oozed from damaged frames inside the hive. This is normally a sign that the bee colony has reached the colony collapse level or the colony has succumbed to major beetles activities. "Leaking" is the term that is commonly used to identify this beetle damage stage. Schafer and Ritter (2014) reported that colonies that are heavily-beetle-infested have a characteristic rotten odor as a result of fermented honey. I have not found this to be a reliable indicator of heavily-beetle-infested colonies.

Fig. 33. To detect small hive beetles in a hive super, place it on an overturned telescoping hive top in a sunny spot for about 10 minutes (a). The bright light will drive the adult beetles down to the bottom. When the super is lifted, adult beetles, if present, will be apparent on the underside of the hive top (b). Use this practice with caution during periods of dearth. Source: Chris Bryan.

Fig. 34. Collapsed honey bee colony showing signs of "leaking" caused by major small hive beetle damage. Hood photo.

Immediate hive removal and treatment of the soil left behind is recommended. If the soil is not treated, a large number of beetles may successfully pupate and emerge as adults to infest other bee colonies.

Genetic Control

Scientists have discovered that African worker bees readily remove unprotected small hive beetle eggs and larvae. This behavioral trait likely plays an important role in the apparent resistance of African bees to beetle infestation. Cape honey bees which only live in the southern tip of Africa have shown the ability to identify capped bee brood cells that the female adult beetles have made a slit and oviposited their eggs. The bees tear into the cells and remove the cell contents including beetle eggs and larvae. These traits likely occur in our European bees at a much reduced level, however these hygienic behavioral traits may possibly be incorporated into a selection program.

Bees often use prisons constructed of propolis to confine adult beetles. African bees are known to collect and utilize more propolis than other bee races, therefore this activity is another possible reason that African bees show resistance to small hive beetles. Selection of bees that utilize more propolis may contribute to small hive beetle resistance.

Fig. 35. Several small hive beetles confined in a propolis prison in corner of hive cover. Hood photo.

Fig. 36. Small hive beetle begging for food from a worker honey bee. This activity is also termed "trophallaxis." Hood photo.

Fig. 37. Several small hive beetles confined in a prison lined with propolis. Hood photo.

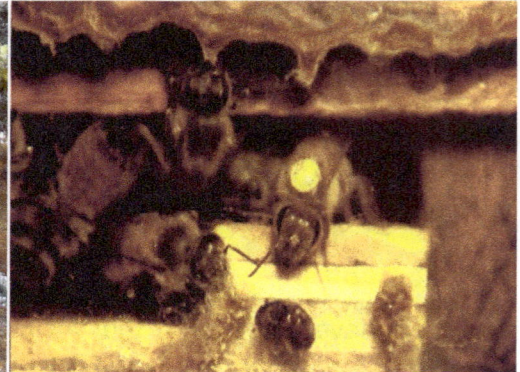

Fig. 38. Adult honey bees guarding a small hive beetle adult. Source: J. Ellis, University of Florida, USA. Bugwood.org

Mechanical Control

Several mechanical trapping devices have been developed in the US and Australia to control small hive beetles. Most of these beetle traps use either vegetable or mineral oil as the beetle killing agent. Caution should be used in the use of these oils because they can also be deadly to honey bees. After use, these oils should be recycled or disposed of properly to prevent environmental contamination.

Small hive beetle traps should play a major role in the integrated management of this hive pest because of their safety in providing control without fear of hive product contamination. Traps normally provide a low cost form of sustained beetle control as long as there is little chance of mass beetle immigration into the apiary. The major disadvantage of most beetle traps is regular trap service is necessary.

Several small hive beetle trap investigations were begun in the US starting in 1998. Plastic bucket traps containing pollen, honey, bee brood, and live bees were placed in apiaries to investigate their effectiveness in trapping adult beetles before they enter honey bee colonies (Elzen et al. 1999). Although some beetles were captured in the bucket traps, the traps proved to be little competition for the more attractive odors coming from managed honey bee colonies. Therefore, most small hive beetle trapping research began to focus on development of an effective in-hive trap. There is still hope that an attractant will soon be developed that is more attractive to flying adult beetles than the odors that are produced by a colony of honey bees (Neumann et al. 2016b).

Fig. 39. Small hive beetle bucket trap investigations in coastal Georgia, USA near Savannah in 1998. Hood photo.

Capturing and killing small hive beetles prior to entry into beehives certainly has many advantages. Traps placed on the perimeter of apiaries could be used throughout the peak beetle movement season to provide a safe and more efficient beetle control option (Annand 2011). Such traps would also be helpful in safeguarding wild or feral honey bee colonies as well bumble bees and stingless bees (Neumann et al. 2016b).

Fig. 40. Small hive beetle pipe traps. Source: Lilia de Guzman, USDA ARS.

Another beetle trap has been developed for use outside hives, but it has been used only to monitor beetle movement into an area (Arbogast et al. 2007). The trap was made of a 25.5 cm (10 inches) section of black PVC pipe with 7.5 cm (3 inches) interior diameter with both ends of the pipe covered with 18-mesh screen cones. A bait made of pollen dough conditioned by allowing male small hive beetles to feed on it for 3 days was placed inside the pipe which was suspended about 1 meter (39.4 inches) above ground. The traps were found to be attractive to beetles preferably when the traps were placed in shade. Few beetles were captured in the traps when placed in full sun. These traps are not marketed and may not compete well with bee colonies for their attractancy.

Efforts have been made to develop hive entrance traps that capture mature small hive beetle larvae as they attempt to drop to the ground to pupate (Arbogast et al. 2012). These traps have been used primarily as a research tool (Neumann et al. 2013) because damage to colonies has already occurred prior to the time when wandering beetle larvae attempt to exit the colony to pupate in the soil. However, an effective beetle larval trap could contribute toward breaking the life cycle of the beetle which is comparable to the purpose of the use of soil treatments.

The use of a moat partially filled with a killing agent such as vegetable oil or soapy water could be used in a similar fashion to kill exiting beetle larvae, if the hive is placed on a stand inside the moat. Rainwater could compromise the system and some returning foraging bees may fall into the moat and die which are some concerns of this idea. This moat trapping procedure may help with controlling beetles for a beekeeper who has only a few hives, but I'm not aware of any research on this measure of beetle control.

A "jar-bottom board small hive beetle trap" was designed and investigated at Clemson University (Hood 2006). The trap consisted of a 2.5 lb. (1.15 kg) square glass honey jar with lid secured by three screws underneath a beehive wood bottom board. The jar exterior was painted black to simulate the dark conditions inside a beehive. A 1.5 inch (3.8 cm) hole was drilled through the hive bottom board and jar lid. The hole was positioned in the center of the bottom board and 5.5 inches (14 cm) from the back of the hive bottom. A screen funnel ("bee escape cone," Betterbee Inc., Greenwich, New York, US) was secured by staples to the lower rim of the hole, allowing the small end of the funnel to protrude down into the jar to impede beetle escape. A 4 x 4 inch (10 x 10 cm) piece of corrugated plastic was secured with staples over the hole on the hive bottom board to impede honey bee entry but provide beetle harborage and entry into the jar below. The jar was filled one-third with apple cider vinegar as a beetle-attractant. From preliminary tests, this trap gave marginal small hive beetle control.

Fig. 41. Black painted 2.5 lb. (1.15 kg) glass jar fastened under hive bottom board. Hood photo.

Most inside-hive small hive beetle traps are based on the fact that adult beetles are constantly searching for a hiding place that is free of adult bee activity. The object of the trap is to provide a small opening large enough for beetle entry but too small to allow for passage of adult bees. A beetle "attractant agent" placed inside the trap provides a bonus chance of attracting more beetles. However, most inhive traps provide no attractancy other than providing harborage from the bees. Most traps do provide a killing agent which beetles come in contact and die.

The West Beetle Trap™ was the first beetle trap marketed in the US. The West Beetle Trap™ is a hive bottom trap that includes a removable plastic tray partially filled with vegetable oil that beetles enter and die. A slotted cover fits tightly over the tray which prevents bees from entering. The trap was initially designed to be serviced through the hive entrance which can be disruptive to the colony. A modified version of the West Beetle Trap™ is now available which can be serviced from the hive rear. A similar modified hive bottom trap known as the Freeman Trap™ has been developed which also utilizes a removable plastic tray with vegetable oil. The Freeman Trap™ is conveniently serviced from the back of the hive. The Freeman Trap™ comes with a screened hive bottom that allows beetles to enter the tray and excludes bees. The Freeman Trap™ is available in varying sizes that fit ten or eight frame hives. Go to www.freemanbeetletrap.com for more information. A "screened bottom with removable rear tray" known as the Beetle Jail Trap™ is available for ten, eight, or five frame hives. Go to www.beetlejail.com for additional information.

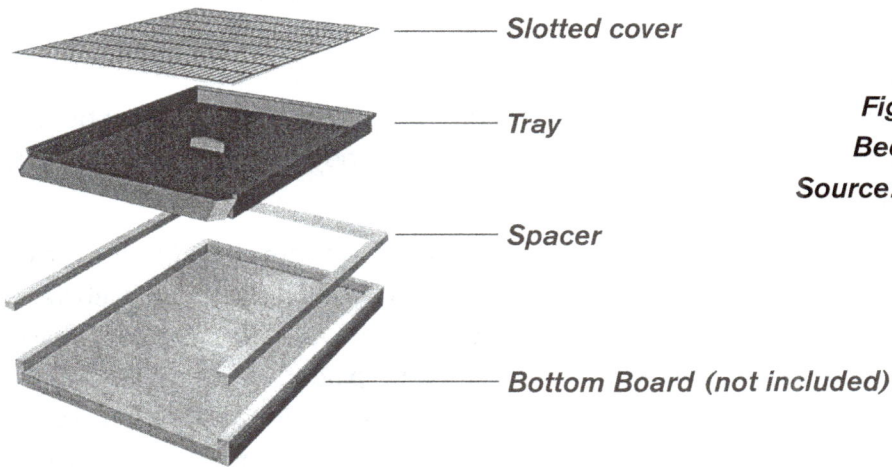

Slotted cover

Tray

Spacer

Bottom Board (not included)

Fig. 42. West Beetle Trap™. Source: Dadant & Sons, Inc.

Fig. 43. Freeman Beetle Trap™ partially filled with vegetable oil removed from hive rear. Hood photo.

Fig. 44. Beetle Jail Bottom Trap™. Source: David Miller

One major advantage of these bottom hive traps that incorporate the use of oils (mineral or vegetable oil) as the killing agent is that they also kill varroa mites, ants, and wax moth larvae which fall into the oil tray. However, oil can be messy to work with and can kill bees if they enter the tray. A tight fitting tray is a must and the beehive should be placed on level ground to prevent oil from overflowing from the tray. In some areas where animals are present such as skunks, the removable tray with vegetable oil must be secured to prevent animal-intervention and possible tray destruction.

Our research at Clemson University found these bottom board traps (Freeman Trap™) with oil trays to be very effective at trapping small hive beetles during the hot summer months (July and August) in South Carolina, USA, when beetle numbers were high (Peterson 2012). The oil trays were found to be a bit messy and sometime trays became propolized and difficult to remove, perhaps also due to warpage of the bottom board materials. The bottom board traps must be constructed precisely to prevent honey bee entry into the oil traps which is lethal to them also.

The Hood Beetle Trap was developed at Clemson University and is a plastic box trap with three separate compartments that can be partially filled with various lethal agents and attractants. The best readily available attractant that I have found is apple cider vinegar which should be placed in the middle compartment and the two side compartments should be half-filled with food grade mineral oil. The trap should be secured inside an empty hive frame and placed in frame position #1 or #10. Beetles enter the one-way beetle trap and become immobilized in the mineral oil and die. The traps should be serviced at roughly 2 week intervals to remove dead beetles and refill the middle compartment with fresh vinegar. During extremely hot and dry summer months, we found the vinegar to evaporate sooner, so refilling more often may be necessary.

Our research with the Hood Trap indicated that roughly the same number of beetles can be trapped in the top super as can be trapped in the brood chamber (Nolan 2008, Nolan and Hood 2010). Trapping in the top super has the advantage of not having to disturb the lower brood chamber when servicing traps and less labor is required. Trapping in the top super only could serve as a small hive beetle monitoring tool as a good indicator of the presence or absence of beetles in the hive. However, trapping in both the brood chamber and the supers above is recommended as a beetle control measure.

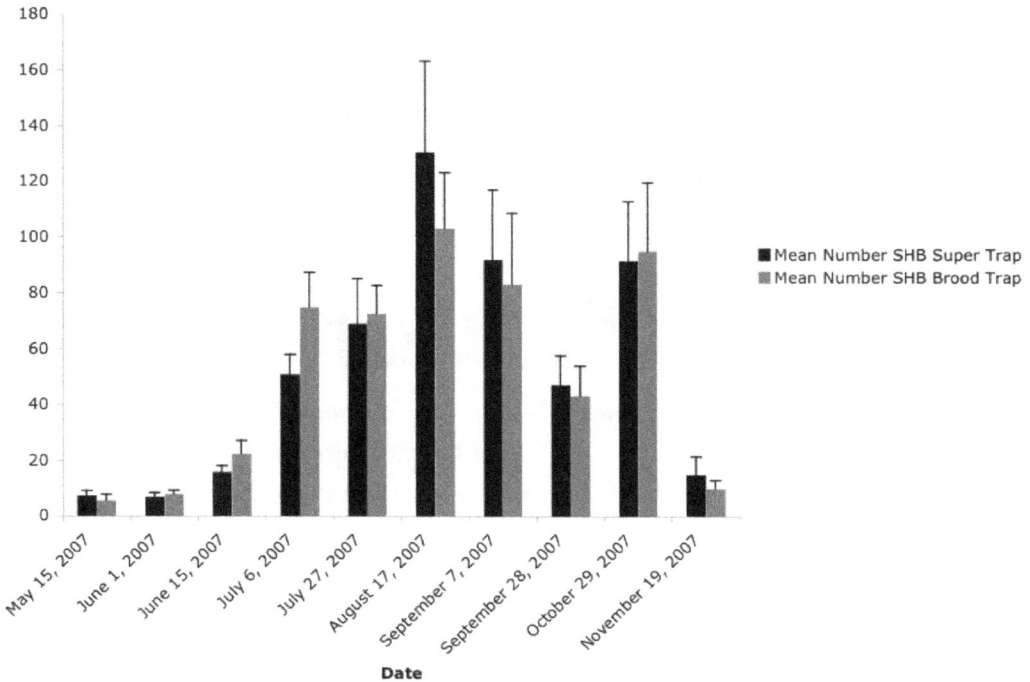

Fig. 45. Mean number of small hive beetle adults caught in Hood Beetle Traps (two Hood Traps per hive; one in brood chamber and one in top honey super) during 10 sampling periods in Oconee and Pickens Counties, South Carolina, USA, May-November 2007. (Five apiaries at ≥ 2.4 km [1.5 mile] distance between apiaries, five colonies per apiary). No significant differences were found over the 7 month trapping period (P > 0.05). (Nolan 2008; Nolan and Hood 2010).

The placement of the Hood Trap in the hive body (brood chamber) has an added advantage of doubling as a drone brood/varroa mite trap. Bees will construct only drone brood cells in the void area of the frame, if placed in frame position #1 or #10 and queens will lay only drone eggs in the cells. When the brood is about 2/3 capped, the beekeeper should cut the comb out and place it in a freezer to kill the varroa which were attracted to the drone brood. If the beekeeper forgets to cut out the capped drone bee pupae and the brood successfully emerges, the resulting varroa population will increase. One disadvantage to the Hood Trap is the bees will sometimes propolize the trap entrance, however a hive tool can be used to quickly remove the propolis.

Fig. 46. Hood Trap mounted on a hive body frame. Hood photo.

Fig. 47. Capped drone bee pupae cut from around a Hood Trap on a hive body frame to be placed in a freezer to kill varroa mites. Hood photo.

Fig. 48. Hood Beetle Trap mounted on shallow honey super frame with middle compartment filled with apple cider vinegar and two side compartments half-filled with food-grade mineral oil Bees will construct comb in the voids in frames placed in honey supers which normally produces some nice cut comb honey. Hood photo.

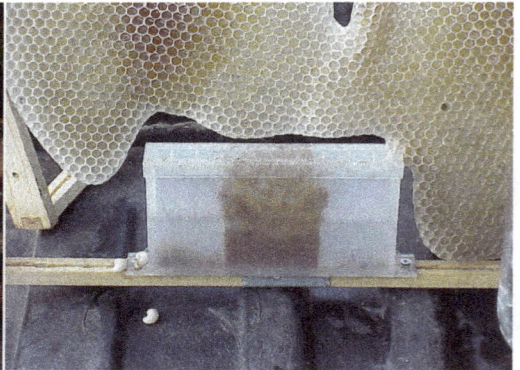

Fig. 49. Hood Trap with middle compartment filled with pollen dough inoculated with **Kodamaea ohmeri** yeast and two side compartments filled with food grade mineral oil. Hood photo.

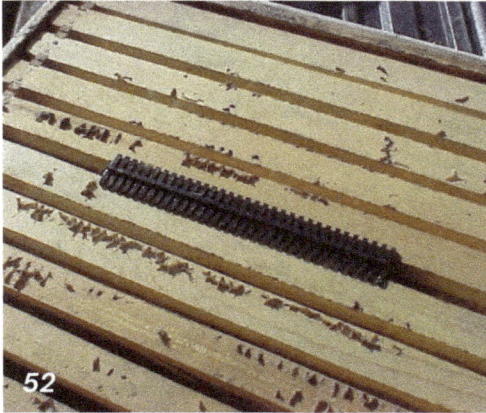

Fig. 50. Beetle Jail Entrance Trap™. Source: David Miller

Fig. 51. AJ's Beetle Eater™ Trap. Source: www.ent.uga.edu

Fig. 52. AJ's Beetle Eater™ trap placed between two hive frame top bars. Hood Photo.

Fig. 53 Beetle Blaster™ trap placed between two frame top bars. Hood photo.

Fig. 54. Beetle Blaster™ trap with several small hive beetle adults captured in vegetable oil. Hood photo.

A beehive entrance trap has been developed for catching small hive beetle adults before they enter the colony. The Beetle Jail Entrance Trap™ is supplied with a removable 2 inch (5 cm) deep molded polypropylene reservoir that can be partially filled with vegetable oil or diatomaceous earth. The reservoir can be removed and serviced easily without opening the beehive. The trap is available in various sizes to fit a Langstroth 10-frame, 8-frame, or 5-frame beehive and can be used also with a top bar or Warre beehive. I have no experience in the use of this trap. For more information on this trap, go to wwwbeetlejail.com.

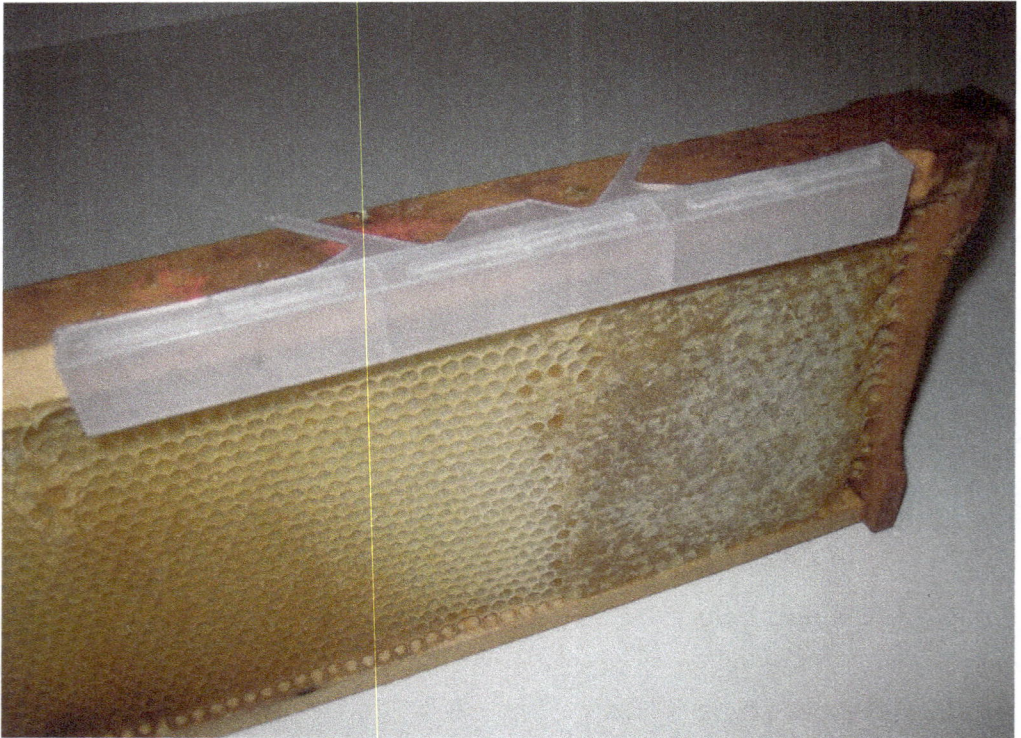

Fig. 55. Beetle Jail Jr.™ plastic trap with three compartments fastened to a hive frame. Source: David Miller.

Fig. 56. Beetle Jail Jr.™ showing trapped adult beetles in oil. Source: David Miller.

AJs Beetle Eater™ trap was developed by an Australian beekeeper and is marketed in the US and Australia. The trap is a two piece longitudinal plastic trap that should be partially filled with vegetable oil and suspended between two frame top bars. Laurence Cutts, former Florida State Apiarist, has developed a similar beetle trap, the Beetle Blaster™. This disposable plastic trap is also designed to be placed between two frame top bars and should be half-filled with vegetable oil. Go to beetleblaster.com for more information. The Beetle Blaster™ and AJs Beetle Eater™ traps can be placed between two frame top bars in the bottom brood chamber or supers above or both.

Another plastic beetle trap for placement between two frames top bars is named the Beetle Jail Jr.™ developed by David Miller. The trap has three compartments and should be filled 1/3 to 1/2 with vegetable or mineral oil with a small amount of vinegar to attract beetles. The trap has a support arm which is used to support the trap by an adjacent frame.

Peterson (2012) conducted research comparing the effectiveness of three small hive beetle traps (Hood trap, Freeman Trap™ and the Cutts Trap (Beetle Blaster™) for removing adult beetles over a 7 months trial period. The Freeman Trap™ proved to be the most effective trap at capturing small hive beetles compared to the other two traps and the controls. More adult beetles were consistently trapped in the Freeman Trap over the 7 months period, and the Cutts Trap and the Hood Trap captured more beetles in late summer than earlier in the year (Peterson 2012).

Sum of Trapped Adult Small Hive Beetles at 2-Week Intervals

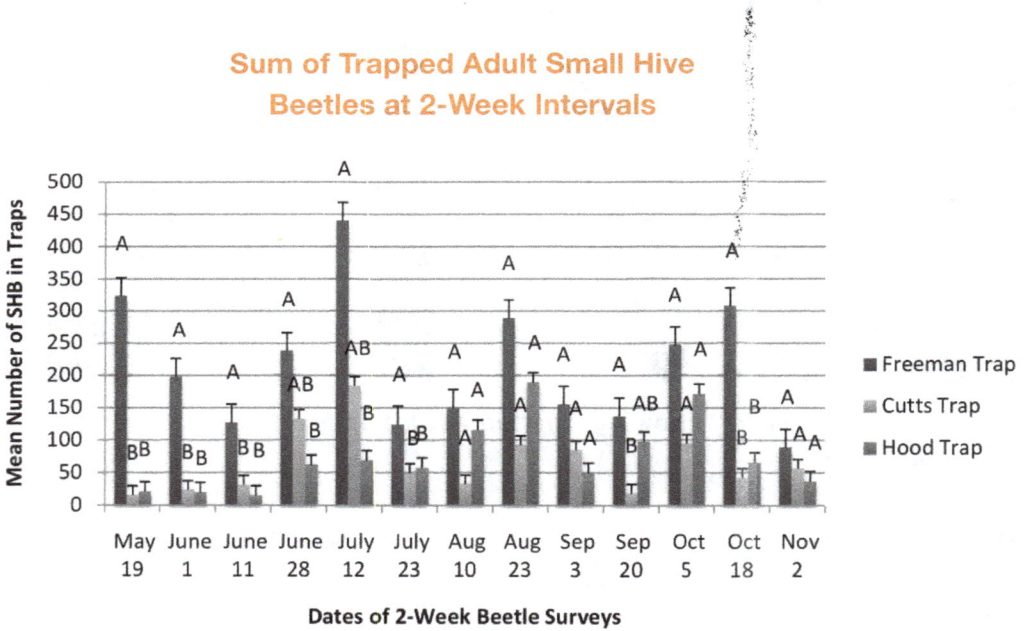

Fig. 57. Mean number of adult small hive beetles ± standard error captured in three small hive beetle traps (Freeman Trap™, Cutts Trap also known as the Beetle Blaster™, and Hood Trap) during 13 sampling periods in Pickens County, South Carolina, USA, May-November 2010. Different letters indicate significant difference (P<0.05). (32 colonies total, eight colonies per apiary, ≥ 1.6 km [1 mile] distance between apiaries) (Peterson 2012).

Kodamaea ohmeri. A yeast, *Kodamaea ohmeri*, has been found to be associated with small hive beetles that when growing on pollen produces an odor
blend that is attractive to beetles (Conklin 2012). A small hive beetle trap bait
containing pollen dough, a mixture of pollen and honey that was conditioned
either by beetle adult feeding or by inoculation with the yeast *K. ohmeri*, was
placed in bottom board traps to successfully monitor beetle populations in
Kenya and the US (Torto et al. 2007; 2010). Investigations were conducted
to examine the growth of *K. ohmeri* inside beehives including water activity,
nutrient availability, the presence of small hive beetles, and the presence of
bees (Conklin 2012). Test results indicated that *K. ohmeri* is capable of growing on bee bread (pollen) collected from beehives without the presence of
beetles and that water activity is important to the growth of the yeast on bee
bread. Small hive beetle frass (feces) contains a high quantity of *K. ohmeri*
and is suspected of being the source of the yeast inoculum (Conklin 2012).
There are other naturally occurring species of yeast found in beehives or in
beekeeper-added pollen supplements that are capable of producing beetle
attractive volatiles that attract adult small hive beetles (Conklin 2012). This
implies that the relationship of beetles and *K. ohmeri* may not be an exclusive
one and that beetles may be attracted to other naturally occurring yeast that
grow on pollen inside beehives. Additional research indicated that female
small hive beetle adults initially preferred pollen fermented by *K. ohmeri* as a
preferred oviposition substrate over control pollen (Conklin 2012).

A simple and quick method of trapping small hive beetles without the use of
oils inside the hive is the use of a 6.4 x 56 cm (2.5 x 22 inch) piece of plastic
corrugation with approximately 0.3 cm (1/8 inch) openings placed in the hive
entrance for a day or two. Schafer et al. (2008) recommended leaving the
strips of corrugation inside the hives for at least 48 hours to give the adult
beetles time to find and enter the refuges. The adult beetles enter the corru-
gations for a hiding place and the beekeeper removes the trap and raps the
trap a few times inside a 5 gal. (19 liters) bucket that has a small amount of
oil or some other killing agent in the bottom (Hood 2011). Or, the beekeeper
can place the trap in a clear plastic bag and secure the bag to prevent beetles from escaping (Williams 2015). The bag with beetles inside can then be
placed in a freezer to kill the beetles. Schafer et al. (2008; 2010c) reported
capture efficiency of 30% of the small hive beetle population in bee colonies

when using this trapping measure. Our preliminary-research conducted at Clemson University using this trapping technique yielded varying results. A clean wooden hive bottom board with all debris removed is a must when using this trap because beetles will hide underneath the trap if the trap is not lying flat. This trapping technique using screened bottom boards is probably less effective and is not recommended during cold weather when beetles are likely inside the bee cluster and are not seeking hiding elsewhere. At best, this trapping technique is useful as a monitoring tool and not as a small hive beetle control measure, however the big advantage of the monitoring technique is that the hive does not have to be opened or manipulated.

Fig. 58. Beekeeper carefully inserting plastic corrugation trap into a beehive entrance. Hood photo.

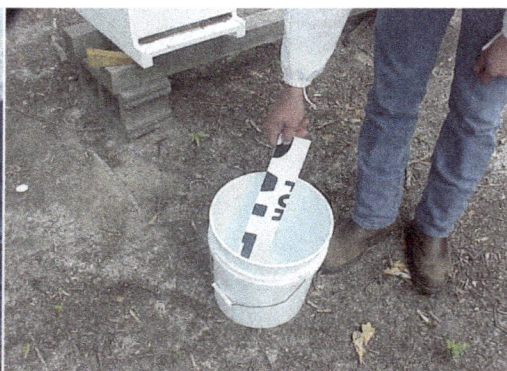

Fig. 59. Rapping the sides of the trap a couple of times on the side of bucket to dislodge the beetles that fall into the oil and die. Hood photo.

Microfiber Towels. The use of microfiber towels or sometimes referred to as kitchen cleaning towels has been suggested as another method of trapping small hive beetle adults (Zacchetti 2015). Beekeepers in Australia and the US report that these towels when placed on top of frame bars become shredded by the worker bees in an attempt to remove them from the hive, but in the process the beetles become entangled in the shredded material and cannot escape (Tew 2015). The suggested theory is that the microfiber strands become ensnared on structures on the small hive beetle legs. This novel approach to trapping small hive beetles has not been supported by research (Tew 2015), however it does appear to be a safe, harmless and economical control technique. Studies need to be conducted to discover the optimum placement of the towels, to investigate the suggested removal and replacement timing, to compare various brands of towels, to observe for any negative effects on the bees and their activities in the hive, and to compare efficiency with other trapping measures.

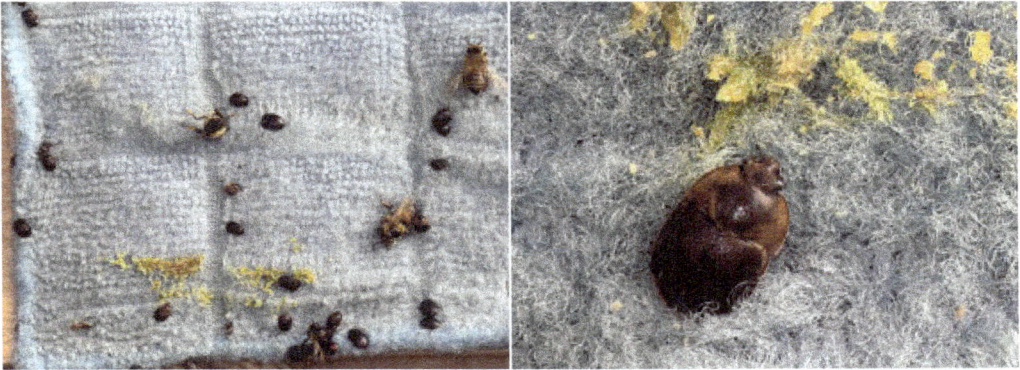

Fig. 60. Small hive beetles trapped in microfibers. Source: Jim Tew.

Attractant Lights and Light Traps. Different light sources have been investigated in an effort to attract adult small hive beetles. Neumann and Elzen (2004) reported poor results when testing white bulb and black light insect traps as beetle attractants. However, UV lights have been used successfully in honey houses to attract wandering larvae. The light should be placed close to the floor in the honey house to attract positive phototactic wandering larvae (Neumann and Elzen 2004; Duehl et al. 2012). The optimum response to beetle larvae and adults was recorded using a UV light with 390 nm wavelength (Duehl et al. 2012). Once the beetles are attracted to the light source, they can be swept up and placed in a container of 50% chlorine bleach/water as the killing agent.

Other research has been conducted to investigate the response of small hive beetles to visual stimuli, testing the effect of the influence of black and white colors and the effect of height on pole traps (de Guzman et al. 2011). White traps captured significantly more beetle adults than black traps likely due to white being more reflective than black. They also reported increased beetle trap catches in pole traps positioned about the same height level as beehive colony entrances in the field (de Guzman et al. 2011). They reported that these traps do not catch high numbers of beetles, as it is clear that honey bee hives attract a majority of the migrating beetle adults.

Fig. 61. Small hive beetle pipe traps. Source: Lilia de Guzman, USDA ARS.

Diatomaceous Earth. Diatomaceous earth has been successfully used in the control of grain beetles, cockroaches, silverfish, mole crickets, and other insect pests. Diatomaceous earth is a powder composed of unicellular/colonial silicified skeletons of diatoms (algae, Bacillarlophyceae) (Cribb et al. 2013). Research has been conducted recently using diatomaceous earth as a possible control agent in small hive beetle traps as a killing agent.

In Australia, laboratory investigations (Cribb et al. 2013) were conducted using diatomaceous earth as a killing agent inside top bar frame traps (Beetle Blaster™). The survival of adult small hive beetles in this study was evaluated in relation to relative humidity (RH=56, 64, 73, 82 and 96%). The lower relative humidity reduced beetle survival which agrees with the hypothesis that diatomaceous treatment results in mortality by water loss from treated insects (Cribb et al. 2013). They reported 100% beetle mortality in treated traps and 8.6% beetle mortality in controls without diatomaceous earth. The diatomaceous earth used during this study was Absorbicide® (Mount Sylvia Diatomite Pty, Ltd, Australia). No field tests were conducted during this investigation.

In research conducted in Australia, South Africa and the US, slaked lime

(CA(OH)^2; also known as hydrated lime), powdered limestone and diatomaceous earth (Fossil Shield® FS 90.0s) were investigated for their effects on small hive beetle pupation and adult emergence in the laboratory (Buchholz et al. 2009). They reported that limestone had no significant effect on beetles. Slaked lime prevented beetle pupation but was lethal only at high dosages of 10 and 15 g per 100 g soil. Low slaked lime dosages 0.5 and 5 g produced > 90% mortality, likely a result of enhanced pathogen activity (Buchholz et al. 2009). Slaked lime and FS 90.0s were also tested in beetle traps (diagnostic trays) in the laboratory and field bee colonies. In field colonies, about 30% of the adult beetles were caught in traps with slaked lime. FS 90.0s resulted in 100% adult beetle morality in field trapped colonies, with about 58% of the adult beetles dying within 48 hours (Buchholz et al. 2009). Additional research is needed to investigate the use of diatomaceous earth and slaked lime, especially in field colonies.

Small hive beetle adults have been suspected to be attracted to the presence and/or activity of other small hive beetles for unknown reasons and possibly from great distances. Neumann and Elzen (2004) hypothesized that beetle pheromone-mediated aggregations might be adaptive to overcome host defenses. There exists the possibility that a pheromone may likely play an important role in small hive beetle long range host finding and aggregations (Neumann et al. 2016b).

Research was conducted at Clemson University, Clemson, South Carolina, USA to investigate the use of a "trap sink" concept to control small hive beetles (Peterson 2012). In other words, are small hive beetles attracted to the odors emanating from the presence and/or activities of other beetles? Individual colonies of honey bees within the same test apiaries were fitted with both one Freeman Beetle Trap™ (tray half-filled with vegetable oil) underneath the brood chamber and one Beetle Blaster™ trap (reservoir half-filled with vegetable oil) in the top honey super to act as "trap sink colonies." Other nearby (distance between apiaries ≥ ¼ mile [.4 km]) test apiaries had no "trap sink colonies" as controls. Data were collected over a 2 year period from April to November of each year. All traps were serviced bi-weekly and beetles counted in traps and all test colonies were sampled for beetle numbers at 6 week intervals. At the end of both test seasons, colonies were dismantled and total beetle counts conducted. No significant difference was found in beetle numbers counted in trap sink colonies vs. non-trap sink colonies within the same apiary or between apiaries having "trap sink colonies" vs. apiaries having no "trap sink colonies". It was observed that extremely low beetle counts were recorded during each of the two test seasons compared to previous years which may have affected the outcome of this experiment. A noticeable lack of rainfall was recorded during the summer months of both test seasons compared to the expected rainfall records in the area. The results of these tests did not confirm the existence of a "trap sink" concept, as tested (Peterson, 2012). More research is needed to investigate the identification and evaluation of the potential impact of various compounds such as aggregation pheromones, food volatiles or any synergism between pheromones and food volatiles on short and long range dispersal and beetle host finding (Neumann et al. 2016b)

Other forms of mechanical hive measures have been investigated that have not proven to provide small hive beetle control. Bottom screens tend to increase hive ventilation and light conditions near the bottom of a hive, but have not proven to increase or decrease the small hive beetle population.

Physical Control

Beekeepers often smash small hive beetles with their hive tools as a form of physical control. If a beekeeper has the time and patience, this activity can reduce the beetle population and contribute to holding the population in check. A small-scale beekeeper from Charleston County, South Carolina, USA once shared with me the fact that he was able to hold his small hive beetle population in check by going through his colonies thoroughly on a weekly basis and smashing adult beetles with his hive tool while his neighboring beekeepers reported losing colonies to beetles.

Battery operated vacuums are also available for beetle removal, however this form of control is for the small-scale beekeeper who only has a few bee colonies. These activities can give the beekeeper a tremendous sense of gratification, but it can be a futile effort when colonies are overrun with beetles and over-manipulation of hive may lead to additional hive problems, such a robbing.

Fig. 62. Hive tool: "beetle smasher." Hood photo.

Fig. 63. Battery operated beetle vacuum. Hood photo.

Fig. 64. Computer micro-vacuum used for alternative beetle control measure. Hood photo.

Concrete Barrier. One form of physical control suggested often by beekeepers is to place beehives on a concrete slab or some other hard surface that is impenetrable by mature larvae as they exit the colony to pupate. The efficacy of this form of beetle control is extremely low because wandering larvae are very mobile and can travel great distances > 200 m (656 feet) (Somerville 2003; Hood 2011). On the other hand, beehives placed on a large city building roof or other large hard surface area may benefit by providing an un-penetrable surface for wandering small hive beetle larvae.

A modified hive entrance in the form of restricting the hive entrance to a single polyvinyl chloride (PVC) pipe has been investigated in an attempt to control small hive beetles. The upper hive entrance did not prove to reduce the small hive beetle numbers in one investigation (Hood and Miller 2005) and two other investigations (Ellis et al. 2002d, 2003) reported inconsistent results. These investigations reported a reduction in bee brood production which would also negate this integrated approach.

Fig. 65. Test apiary with beehives fitted with 3.5 cm or 1.4 inch (interior diameter) upper PVC pipe entrance. Hood photo.

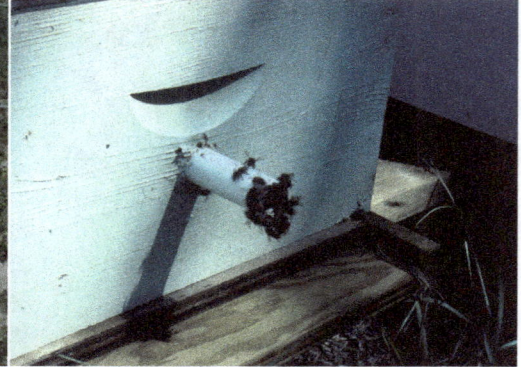

Fig. 66. Beehive fitted with a 1.9 cm or ¾ inch PVC pipe hive entrance. Source: J. Ellis, University of Florida, USA.

In some of our small hive beetle research, we have used vacuums and aspirators to remove beetles from colonies in the fall (before bees begin to cluster) to obtain a total colony beetle count. This is a laborious and time consuming task that requires a minimum of two people, but may pay big dividends for the small-scale beekeeper to reduce the number of overwintering beetles. The procedure begins by finding and placing the queen in a cage for safe keeping. Then each hive frame is removed and shaken on an 8 x 3 feet (2.4 x .9 meter) white plastic table to free bees and beetles. Next, the frame top bar edge is lightly bounced a couple of times on the table top to free any remaining beetles that are hiding in cells. The frame is then turned over and the frame top bar bounced again on the table top to remove any beetles from the other side. One person manipulates the frames as another person stands on the opposite side of the table and vacuums or uses an aspirator to collect the beetles from the table top, counts beetles and brushes bees to the side. The boxes, bottom, and hive top should also be bounced on the table to remove and capture remaining beetles. After all the equipment has been processed in this manner, the frames are reloaded into the hive, queen released, and bees remaining on the table brushed back into the hive. For research purposes, we released the captured beetles back into the hive to continue the project.

Fig. 67. Shaking honey bees from frames and vacuuming or aspirating beetles. Hood photos.

This radical technique is no doubt very stressful to a colony, but has proven to remove at least 80% of the beetles, as reported by scientists who have used this approach. A few beetles will get by undetected and a few will fly away safely and return to a colony. This technique has been used to only count beetle numbers in bee colonies and its effectiveness as a control tool has not been investigated. As a beetle control technique, simply smashing beetles with a hive tool will likely be preferred as opposed to safely removing the beetles. The beekeeper can expect to kill a few honey bees during the process of eliminating the beetles.

Another physical control technique is to move beetle-infested colonies to a new location. Some US commercial or migratory beekeepers now report having fewer beetle problems, as long as they keep their colonies on the move. Moving colonies simply breaks the beetle life cycle by leaving the mature larvae and pupae behind in the soil. In some of my field research, I have noticed that many beetle adults and larvae will exit a hive during transportation, perhaps responding to vibrations. I have often noticed great numbers of beetle larvae and adults left behind in my pickup truck bed following the movement of beetle infested colonies or deadouts. Further investigations into this phenomenon may lead to a form of control of this pest.

Leaving colonies in the same apiary where beetles have been a major problem for years is not recommended. The recommended minimum distance required to move beetle-infested colonies to a new location has not been investigated.

Fig. 68. Moving honey bee colonies from an apiary known for having a history of small hive beetle problems to a new location is highly recommended. Photo Source: Thien Gretchen, Bee Ranch, Texas, USA.

If it is evident that several hundred adult beetles are present in a colony and beetle larvae are present in a colony resulting in possible honey fermentation, the entire hive should be removed from the apiary and treated in a remote location. Annand (2011) recommended the removal of susceptible hives from an apiary to reduce opportunities for further beetle reproduction and to help minimize population expansion within the same apiary. An alternative option is to place the hive and its contents into a freezer for a couple of days which will kill all beetle life stages. Regardless, the entire hive should be removed from the apiary before more larvae exit the hive to pupate in the soil. Remember to treat the soil left behind with a soil treatment to kill any beetle pupae before they emerge as adults.

Anything that reduces the ratio of bees-to-comb surface when beetles are present can lead to major beetle problems. Over-supering and swarming are two examples that can result in increased beetle problems, as well as wax moth or skunk problems.

In areas where beetles are problematic, beekeepers should not use a Porter bee escape to remove the honey crop. Honey supers left above a Porter bee escape for more than a day or two stand a high chance of destruction by beetles which thrive in warm locations that are free of bees. Pollen traps should also be serviced regularly and maintained carefully because the collected pollen serves as unprotected source of protein which can enhance beetle reproduction.

Freezing a couple individual frames that contain a few beetle larvae from a live bee colony has been investigated, but this will rarely result in successfully salvaging a colony that also shows signs of weakness and low morale. A close examination of these beetle larvae infested frames will often reveal wax moth larvae too. Two measures that may help increase the chance of success are: 1) to replace the beetle larvae-infested frames with frames filled with brood and hanging bees from other healthy colonies that show a high bee-to-comb ratio to boost the bee colony population or 2) to move the remaining beetle-free frames down to a nucleus-size (five-frame) box where the bees can better cover and protect the frames. This is a very risky beetle control measure that may be successful only when other colonies in the same apiary appear to have low beetle populations.

When honey-filled supers are removed from colonies that are beetle-infested, it is highly recommended to extract the honey within 2 days. However, if this is not possible, the beekeeper is advised to maintain a relative humidity of < 50% inside the honey house. An additional method of controlling beetles in supers of honey was recommended by Annand (2011) by placement of supers in cold storage of temperatures below 15° C (59°F). He also recommended the option of maintaining low humidity in the honey house, but his recommendation was to keep the relative humidity less than 34%. The low humidity results in desiccation of beetle eggs and larvae that were transported into the honey house inside the honey supers. Beetle larvae can cause complete loss of the honey crop inside the honey house, if these guidelines are not followed. Frames of honey which have been used in the past as brood frames are more vulnerable to beetle reproduction because combs contain small amounts of pollen and brood left behind. Pollen and bee brood provide small hive beetles the needed protein for development and reproduction. Wet supers that honey has been extracted should be removed also from the honey house immediately because the residual honey contains pollen.

Beekeepers are also advised to practice good sanitation around the honey house to avoid beetle problems. Timely removal of bits of comb, wax cappings, and pollen is highly recommended because these materials are highly attractive to beetles and may provide resources needed by beetles for reproduction.

A 50% bleach/water solution has been shown to kill beetle larvae in honey houses and for use in cleaning or salvaging larvae-infested comb after 4 hours of treatment (Park et al. 2002). Treated comb should be set aside for at least 24 hours to allow the bleach odor to dissipate.

Feeding honey bee colonies pollen substitute patties was reported to be problematic because increased beetle reproduction occurred shortly following placement of the patties (Westervelt et al. 2001). Research was conducted at Clemson University to investigate the effects of feeding bees pollen substitute patties (Global Patties, Bay 2-8 Eastlake Way NE, Airdrie AB, Canada T4a 2J3) in mid-winter (Hood 2009b). This activity is practiced by some beekeepers to supplement colony nutrition, particularly when little pollen is available in the beehive. We discovered that female small hive beetles are capable of laying eggs in the patties which were located in the warm area just above the honey bee cluster. The eggs hatched and beetle larvae were found primarily in the patties. However, the beetle larvae were unable to survive when leaving the warm area of the hive as many mature dead larvae were found on the bottom board in their attempt to exit the hive to pupate. The results of these investigations indicate that there is low risk when beekeepers feed pollen substitute patties in the winter when conditions are unfavorable for beetle development. However, beekeepers should be conservative in feeding pollen substitute patties when small hive beetles are present in fall or early spring when mild temperatures may persist and result in successful beetle reproduction (Hood 2009b).

Fig. 69. Pollen substitute patty placed over the bee cluster in winter. Hood photo.

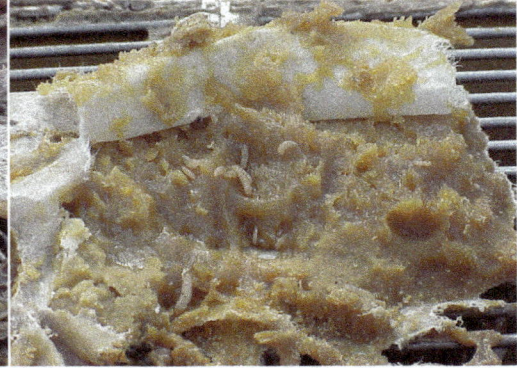

Fig. 70. Notice small hive beetle larvae in patty. Hood photo.

Biological Control

Research investigations have been conducted to find an effective form of biological control for small hive beetles (Ellis et al. 2010; Shapiro-Ilan et al. 2010). Infectivity tests under field conditions found that small hive beetle larvae were susceptible to soil infesting entomopathogenic nematodes (*Steinernema riobrave* or *Heterorhabditis indica*), but field tests have yet to confirm their sustained reliability in the field. Soil nematodes, *Heterorhabditus indica*, are available for purchase from Southeastern Insectaries, Inc., Perry, Georgia, USA, ph. 877-967-6777 or email: sei@aaltel.net. Research conducted in Mississippi, USA, reported a natural infestation of unidentified species of nematodes infesting small hive beetle adults that were collected from soil samples (de Guzman et al. 2009). Entomopathogenic nematodes have the potential to be used as biological control agents against small hive beetles, but more research is needed to verify their reliability under varying conditions. Timing will be critical in soil nematode treatments because effective application will have to be coordinated with beetle larval migration or pupation (Pettis and Shiminuki 2000).

Three strains of *Bacillus thuringiensis* Berliner (*B. thuringiensis* var. *aizawai*, *B. thuringiensis* var. *kurstaki* and *B. thuringiensis* var. San Diego *tenebrionis*) were investigated and showed no ill effects on the reproductive success of small hive beetles (Buchholz et al. 2006).

Fig. 71. Beekeeper applying soil nematodes to control small hive beetles during a research project at Clemson University, South Carolina, USA. Hood photo.

The ability of entomopathogenic fungi to kill small hive beetles was first reported by Lundie (1940). Various fungi have been isolated from/or used effectively against small hive beetles (Richards et al. 2005; Cabanillas and Elzen 2006; Muerrle et al. 2006; Ellis et al. 2010). An infectious fungus (*Aspergillus flavus*) has been identified that infects small hive beetles, however the utilization of the fungus for beetle control has not proven to be safe because of its side effects on bees and fear of honey contamination (Ellis 2004). Various other fungi have been isolated from small hive beetles or have been proven to be effective in controlling beetles (Richards et al. 2005; Muerrle et al. 2006). Leemon and McMahon (2009) and Leemon (2012) have reported good efficacy against small hive beetles in the laboratory and the field assays using isolates of *Metarhizium* and *Beauveria*. *Beauveria* isolates performed best against adult beetles and *Metarhizium* isolates performed best against beetle larvae.

Imported fire ants (*Solenopsis invicta*) which are found throughout the southern US feed on soil infesting insects and likely feed on mature small hive beetle larvae when they enter the soil to pupate. Fire ants are opportunists and may play a role in conjunction with other IPM tools, but they have not been found to be relied upon as a stand-alone beetle control option, even when ant mounds are present in or near the apiary (personal observations). Pesticide soil treatment such as permethrins that are used to kill pupating small hive beetles will also kill fire ants which will negate any beneficial effects of this insect.

Scientists have isolated a beneficial yeast, *Kodamaea ohmeri*, which produces volatile profiles that are attractive to small hive beetles and contains compounds also found in the honey bee alarm pheromone (Benda et al. 2008). Lundie (1940) first described this slimy material that he found on combs infested with small hive beetle larvae. A dough containing the yeast has been field tested to investigate the usefulness of the yeast as an attractant in adult beetle traps (Arbogast et al. 2007; Torto et al. 2007; Nolan and Hood 2008; Hayes et al. 2015). Additional research is needed to investigate the usefulness of this yeast in the management of small hive beetles. In addition to being a possible attractant to small hive beetles, Neumann et al. (2016b) suspects a symbiotic relationship exists between the beetles and *K. ohmeri*, whereby contaminating the honey with the yeast that allows the yeast to grow. The process results in honey fermentation that may provide protein thereby supplementing the beetle larval diet when feeding predominantly on honey (Neumann et al. 2016b). Schafer et al. (2009) reported that some of the organic acids used for varroa mite control inhibit the growth of *K. ohmeri*. Santino et al. (2013) indicated that there may be a human health risk associated with the use of *K. ohmeri*.

Adult small hive beetle aggregation pheromones may play a role in overcoming honey bee host defenses (Neumann and Elzen 2004). Laboratory studies using plastic cages and field apiary investigations have indicated that beetle distribution is different from a random distribution (Mustafa, et al. 2006). Neumann and Elzen (2004) reported low beetle infestations in an apiary containing several African bee (*Apis mellifera scutellata*) colonies except for one colony that contained 491 adult beetles. Spiewok et al. (2007) noted that adult beetle infestation levels were not significantly influenced by colony phenotypes (number of bees, amount of brood, or honey stores) thereby indicating that cues other than colony size or food stores are responsible for their attractiveness (Neumann et al. 2016b). Control programs for other Nitidulidae species include the use of aggregation pheromones (Petroski et al. 1994; James et al. 2000), therefore similar pheromones may play a major role in long-range host colony finding and aggregations of the small hive beetle (Neumann et al. 2016b). A synergistic effect between host food odors and aggregation pheromones may play a role in attracting small hive beetles as is the case of another Nitidulid species *Carpophilus lugubris* (Lin et al. 1992). Research is needed to investigate the possible discovery and utilization of aggregation pheromones in small hive beetle management programs.

Small hive beetle mating disruption is another form of biological control of this pest that has been noted in the literature. In her initial data recordings, Conklin (2012) noted that beetles may require an extended mating period to become fertile which may provide an opportunity for mating disruption to control of the small hive beetle. Downey et al. (2015) and Neumann et al. (2016a) conducted research on laboratory reared small hive beetle males that were radiated for sterilization and released. They reported that doses required for male sterilization (>75 Gy) resulted in 100% mortality after 7 days (Neumann et al. 2016a). A high level of sterility (99%) was achieved when irradiating pre-reproductive adults of both sexes at 45 Gy under low oxygen (1-4%) while reporting moderate survivorship which may suffice for sterile insect releases (Downey et al. 2015). Additional research in small hive beetle mating disruption techniques is needed, however Mustafa (2015) noted that the beetle mating system may reduce the chances of success.

Chemical Control

Chemicals when used inside beehives for pest control should be applied as a last resort, however beekeepers often use pesticides as a first choice when confronted with a new pest like small hive beetles because of their convenience, ease of use and often immediate results. But, there are some serious disadvantages to the use of pesticides inside beehives including possible honey and comb contamination, such as the discovery of coumaphos residues found in beeswax (Mullin et al. 2010). Another possible issue is beekeeper dependence on pesticides to control a pest when their use can be short-lived due to development of pest resistance to the chemical, especially like small hive beetles due to their high mobility and fecundity (de Guzman et al. 2011).

Several chemicals have been tested and some registered for small hive beetle control in countries where small hive beetles are found. In the US, laboratory investigations were conducted on 14 insecticides and four insect growth regulators for their effect on small hive beetles and their results indicated that beetles were susceptible to several classes of insecticides (Kanga and Somorin 2011). The LC50 (lethal concentration of 50% mortality) to small hive beetle adults was 0.53, 0.53, and 0.54 µg/vial for fenitrothion, chlorpyrifos, and methomyl, respectively. However, against the beetle larval stage, fenitrothion was the most toxic at an LC50 of 0.89 µg/vial compared to chlorpyrifos at an LC50 of 1.64 µg/vial which was similar to the LC50 of 1.21 µg/vial for fluvalinate and 2.24 μg/vial for methomyl (Kanga and Somorin 2011). Surprisingly, these pesticides were found to be more toxic to small hive beetles than the organophosphate coumaphos (LC50 of 1.61 μg/vial and 1.32 μg/vial for beetle adults and larvae, respectively) found in Check Mite+®, which is currently registered for small hive beetle control in the US. The insect growth regulators tested did not show promise in controlling all larval stages of the small hive beetle (Kanga and Somorin 2011). Although these investigations may provide useful insights and baseline data for future development of small hive beetle control strategies in honey bee colonies, beekeepers should resist the temptation of using non-registered pesticides. Additional research will be required to investigate the possible harmful effects of the chemicals on honey bee exposure and to assure the materials meet safety requirements for human consumption of honey and other hive products.

Other chemicals have been investigated for small hive beetle control including formic and acetic acid. Although Amrine and Noel (2006) reported that formic acid treatments failed to control small hive beetles in their test colonies, Schaffer et al. (2009) concluded that treatment with 60% formic acid had a negative impact on beetle larvae on combs in nucleus hives containing two honey/pollen frames. Further tests conducted with similar nucleus hives using 70% acetic acid treatments resulted in higher beetle adult mortality when compared to control nucleus colonies (Schaffer et al. 2009). However, Buchholz et al. (2011) in further investigations reported that formic acid treatments in field colonies did not significantly increase small hive beetle mortality.

Check Mite+® (Active Ingredients: 10% Coumaphos), the same product initially developed to control *Varroa destructor* was one of the first to be registered in the US for small hive beetle control inside honey bee colonies, but the product can be used legally only during non-nectar flow periods. A single strip of the product should be cut in half and attached underneath a 10 x 10 cm (4 x 4 inch) piece of corrugated plastic or cardboard and placed near the back of the hive on the bottom board. The piece of plastic or cardboard serves as a hiding place or trap and the beetles receive a lethal dose of the pesticide upon contact. Varying results have been reported by beekeepers using this product. This product stands little chance of controlling beetles in late winter, early spring, or late fall when adult beetles are normally in-active or confined to the bee cluster. Beekeepers should carefully use Check

Mite+® only when other forms of control have failed. Beekeepers must follow the pesticides label directions and resist the temptation of using the product during nectar flow periods or when honey supers are on the hive. The product must be removed from the hive in a timely manner, according to the directions. Ellis and Delaplane (2007) reported from their research with Check Mite+® that wandering beetle larvae continued to burrow following 24 hrs exposure to the product.

Fig. 72. Check Mite+® fastened to 4 x 4 inch piece of plastic corrugation. Manufacturer's photo on left and Hood photo on right.

Another use of Check Mite+® in a hive bottom trap is the Beetle Barn™ that has been used by US beekeepers. Bernier et al. (2015) reported good results with this trap using a small piece of Check Mite+® placed in the center of the trap. Again, beekeepers are cautioned to carefully follow labeled directions to prevent honey and other hive product contamination.

Fig. 73. Beetle Barn™. Source: Product Manufacturer.

Another disposable hive floor trap, Apithor™, has been developed, tested, and patented in Australia by Rural Industries Research and Development Corporation and NSW Department of Primary Industries (DPI), after much work by NSW DPI entomologist Garry Levot. The active ingredients (AI., 0.48 g/kg Fipronil) of the chemical used in the trap last up to 3 months. The two-piece rigid plastic shell trap is made to hold a fipronil-treated piece of corrugated cardboard insert (Levot 2008; Levot and Somerville 2012). Levot (2012) reported positive control of small hive beetles in colonies compared to control colonies. He noted there was no ill effect on the honey bees and chemical residues did not exceed 1μg/kg.

Fig. 74. APITHOR™ small hive beetle refuge trap prior to final assembly and sealing. Source: Garry Levot, (Aust. J. Entomol. 51: 198-204).

Fig. 75. APITHOR™ placed on hive floor of beehive. Source: Garry Levot (Aust. J. Entomol. 51: 198-204).

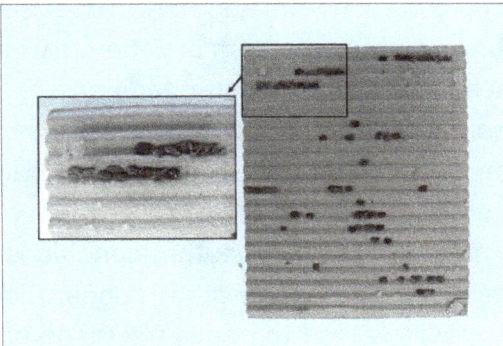

Fig. 76. Dead beetle adults inside a deconstructed harborage (APITHOR™) removed from a hive at the completion of a field trial. Source: Garry Levot, (J. of Apic. Res. 47: 222-227).

Much research was conducted in the developmental stage of the product Apithor™ by scientists to test for possible hive product contamination in laboratory and field trials. Extreme care was made to assure that adult beetles entering the trap died in the trap and did not escape after receiving a possible sub-lethal dose of the chemical. Apithor ™ comes as a ready to use disposable sealed trap product that should not be tampered with by the beekeeper. Beekeepers must not attempt to construct their own version of this trap.

A form of chemical control of all life stages of small hive beetles in stored beekeeping equipment is to fumigate the equipment with aluminum phosphide (phosphine). This is a very toxic chemical (Annand 2008) and extreme caution should be taken by the beekeeper and labeled directions must be followed. In many countries this is a restricted use pesticide and training is required for purchase and use of the product.

The use of soil drench products to kill mature beetle larvae or pupae in the ground need to be applied in all directions from the hive for about 180-360 cm (6-12 feet) to maximize efficacy (Pettis and Shiminuki 2000). Gard Star® (A.I. 40% permethrin EC) is marketed as a soil drench pesticide and is used to kill mature beetle larvae as they exit the hive to pupate in the soil. Care should be taken to avoid spraying this pesticide on the hive entrance which would result in killing honey bees. Gard Star® can also be used to treat the soil underneath dead-out colonies to prevent beetles from emerging and entering other nearby colonies. Since this product is not used inside the hive, there is little chance of hive product contamination. Therefore, the use of this pesticide may be used more freely in an IPM program until we can find a more suitable and efficient biological agent for killing beetles in the soil. From a beetle reproductive control approach, Gard Star® should be used only when beetle larvae are present in the colony. In my experiences in the southeastern US, I have seen very few beetle larvae in colonies in April and May in upstate South Carolina, USA. The months of June and July are normally the time of year when beetle reproduction increases dramatically when conditions are favorable, so beekeepers need to be more vigilant during these two months in South Carolina. However, one problem with the use of Gard Star® is that we simply do not know how long the product remains lethal to beetles in the soil, which is likely dependent on environmental and weather conditions such as temperature, sunlight, soil type and rainfall. The other concern is that widespread overuse of Gard Star® will likely lead to beetle tolerance or

Fig. 77. Gardstar® 40% Permethrin E.C. Soil Drench Product. Source: Product Manufacturer.

resistance to the product in a few years. This is similar to the current problem that we are having with varroa mite resistance to certain pesticide products in the US and other parts of the world.

Other soil treatments for small hive beetle have been investigated in South Africa using HCH (benzene hexachloride, carbaryl, chlordasol, and salt solutions (Cuthbertson et al. 2013). Chlorodasol proved to be the most effective at killing small hive beetles in the soil in their tests. In Australia, the National Registration Authority issued a permit to allow Farmoz Permex EC insecticide plus other registered products containing 500 g/l permethrin to be used as a soil treatment for small hive beetles (White 2003). Mutinelli et al. (2014) noted that pyrethroid products containing cypermethrin and tetramethrin are used as ground drenches in Italy to control small hive beetles.

PRECAUTIONARY STATEMENT. Beekeepers should resist the temptation of using off-brand or unregistered pesticides for small hive beetle control. There are great risks involved when a beekeeper breaks the law (national and state) when using a pesticide that is not registered for its labeled application. The pesticide label is the law and should be followed carefully by beekeepers. We have found that beeswax readily absorbs chemicals and may harbor toxic materials for long periods of time. Using illegal pesticides for small hive beetle control may lead to contaminated

hive products and can result in injury to the consumer as well as the beekeeper. Our beekeeping industry can ill afford the public outcry over the news of pesticide-contaminated honey.

Summary

Prior to 1996, the small hive beetle was known as a hive pest of little consequences mainly in Sub-Saharan Africa. But, over the past 20 years the beetle has now become a global threat to many regions of the world particularly anywhere European honey bees are found. The beetles' future impact will be more pronounced in areas having a warm and humid climate along with soil conditions favorable for small hive beetle reproduction such as the Caribbean, the Mediterranean, Southeast Asia, and South America. However, the small hive beetle has been spread to even some cold climates in Australia, Canada, and the USA which leads one to predict that the beetle has the ability to "hitch a ride" right across the world (Williams 2015).

Currently, the small hive beetle has a range expanse covering six continents and will likely spread to new regions of the world soon. Much of this range expansion was a result of movement within managed honey bee colonies or associated products. Other suspected avenues of range expansion include movement of beetles on fruit or in soil associated with plant trade, however no hard evidence so far exists to substantiate this theory. Much progress has been made in our knowledge of this devastating hive pest over the past 20 years in relation to its behavior, biology and control, but further research is greatly needed to advance our understanding of this hive pest.

From a beekeeping perspective, small hive beetles can be present in a honey bee colony in low numbers and not be a problem. However, beekeepers are advised to monitor their colonies closely and be prepared to take action, especially during certain times of the year when beetle reproduction tends to increase. Beetles do have the ability to reproduce quickly when conditions are favorable and colonies are stressed. There are many recommendations and tools available to the beekeeper to manage this hive pest.

Sometimes when conditions are favorable for small hive beetle immigration and beetle reproduction is high, the beekeeper is in for a real challenge to control this hive pest. Large numbers of beetles have been reported to enter single bee colonies which can overcome the natural defenses of even a strong bee colony. There are a few reports in the literature of migrating swarms of beetles entering a single hive. Fortunately, this occurs very infrequently, so it is up to the beekeeper to help the bees in maintaining low beetle populations by using a combination of safe and effective integrated pest management tools and recommendations. There are now many control options available to beekeepers to manage small hive beetles, however beekeepers are urged to select control measures that will maximize efficiency but at the same time choose those control measures that will have the least environmental impact. In most cases, the integrated management of small hive beetles will serve well to control this hive pest.

Winter is a good time for you to sit back and evaluate how well your beetle management efforts worked last year. Maybe the beetle levels increased to the point of negatively impacting your colonies or perhaps colonies seemed to be overrun by beetles in some apiaries. On the other hand, beetles may have been present, but in very low numbers. Having many tools and recommendations available for you to consider, maybe it is time to try a combination of control options and not depend on a single method.

Top 20 small hive beetle management recommendations

1. Maintain healthy, strong honey bee colonies to promote a high bee-to-comb ratio

2. Place colonies in full sunlight to provide dryer soil conditions which helps reduce successful beetle pupation

3. Monitor colonies for beetle infestation levels particularly during the season when conditions favor beetle reproduction

4. Physically kill or remove beetles when inspecting a colony, but do not leave equipment exposed for long periods of time which may lead to robbing

5. Resist the temptation of over-manipulating beetle-infested colonies which often promotes beetle reproduction

6. Do not over super colonies unless there are plenty of bees to patrol the extra combs

7. Do not use inner covers, end frame spacers, or porter bee escapes as they provide harborage for beetles

8. Treat soil with Gard Star® or a similar registered product around the colony when you first observe beetle larvae inside a hive

9. Be very conservative in feeding colonies pollen substitute patties during favorable weather conditions that promote beetle reproduction

10. Trap beetles using one or more of the trapping devices presently marketed

11. Service pollen traps often because they provide optimum food and protection necessary for beetle reproduction

12. Do not combine beetle-infested honey supers from weak colonies with strong beetle-free colonies

13. If an apiary has a history of beetle problems, move the colonies to a new location

14. Move heavily beetle-infested or dead colonies to another location and treat or sanitize equipment and treat soil left behind after removal

15. Propagate from queens whose colonies show resistance to beetles

16. Use Checkmite+® or other similar registered pesticides inside hives as a last resort

17. Use only pesticides that are registered for small hive beetle control and apply them according to labeled directions

18. Extract honey from supers within 2 days following hive removal

19. Maintain low relative humidity < 50% in the honey house if storing honey supers for more than 2 days.

20. Maintain good sanitary conditions inside and outside the honey house by removing comb, wax cappings, pollen, and slumgum.

Disclaimer statement

Pesticides recommended in this book were registered for the prescribed uses when printed. Pesticide registrations are continuously being reviewed and may be revoked for proper justification. Should registration of a recommended pesticide be canceled it would no longer be recommended by this author.

Use of trade names in this publication is for clarity and information; it does not imply approval of the product to the exclusion of others which may be of similar, suitable quality or composition, nor does it guarantee or warrant the standard of the product.

References

Amrine, J.W., Noel R. (2006) Formic acid fumigator for controlling honey bee mites in beehives, Virginia Agricultural and Forestry Experiment Station, Scientific Article Number 2952 [online] http://www.wvu.edu/~agexten/varroa/FormicAcid.pdf

Annand, N. (2008) Small hive beetle management options, Primefacts. NSW DPI, 764, 1-7 pp.

Annand, N. (2011) Investigations of small hive beetle biology to develop better control options. MSc. Thesis, University of Western Sydney, Australia.

Arbogast, R.T., Torto, B., Engelsdorp, D., Teal, P.E. (2007) An effective trap and bait combination for monitoring the small hive beetle, *Aethina tumida* (Coleoptera: Nitidulidae). Florida Entomologist 90(2): 404-406 pp.

Arbogast, R.T., Torto, B., Teal, P.E. (2010) Potential for population growth of the small hive beetle *Aethina tumida* (Coleoptera: Nitidulidae) on diets of pollen dough and oranges. Fla. Entomol. 93(2): 224-230 pp.

Arbogast, R., Torto, B., Williams, S., Fombong, A., Duehl, A., Teal, P. (2012) Estimating reproductive success of *Aethina tumida* (Coleoptera: Nitidulidae) in honey bee colonies by trapping emerging larvae. Environ. Entomol. 41(1): 152-158 pp.

Arias, H. D. (2004) Small hive beetle infestation (*Aethina tumida*), El Salvador, OEI report.

Benda, N., Boucias, D., Torto, B., Teal, P. (2008) Detection and characterization of *Kodamaea ohmeri* associated with small hive beetle *Aethina tumida* infesting honey bee hives. J. Apic. Res. Bee World 47(3): 194-201 pp.

Bernier, M., Fournier, V., Giovenazzo, P. (2014) Pupal development of *Aethina tumida* (Coleoptera: Nitidulidae) in thermo-hygrometric soil conditions encountered in temperate climates. Apiculture and Social Insects 107(2): 531-537 pp.

Bernier, M., Fournier, V., Eccles, L., Giovenazzo, P. (2015) Control of *Aethina tumida* (Coleoptera: Nitidulidae) using in-hive traps. Can. Entomol. 147(1): 97-108 pp.

Berry, J. (2009) Small hive beetle round-up/Beetles come on strong in the south right now-be ready! Bee Culture: 38-40 pp.

Borror, D., Triplehorn, C., Johnson, N. (1989) An Introduction to the Study of Insects. 6th Edition. Harcourt Brace College Publishers, Ft. Worth, Philadelphia, New York: 438 p.

Borroto, H., Chan, S., Demedio, J. (2014) Diagnostico de *Aethina tumida* Murray (Coleoptera: Nitidulidae) en colmenas (*Apis mellifera* L.) de mayabeque, Memorias Jornadas Cientificas por el 122 Aniversario del Sabio de la Medicina Veterinaria Cubana Dr. Ildefonso Perez Vigueras, Univesidad de Ciencias Medicas – Consejo Cientifico Veterinario Pinar del Rio, Cuba, 2014.

Brion, A.C.B. (2015) Small hive beetle poses threat to bee industry. The Philippine Star, [online] http://www.philstar.com/agriculture/2015/02/22/1426217/small-hive-beetle-poses-threat-bee-industry.

Brown, M.A., Thompson, H.M., Bew, M.H. (2002) Risks to UK beekeeping from the parasitic mite *Tropilaelaps clarae* and the small hive beetle, *Aethina tumida*. Bee World 83(4); 151-164 pp.

Brown, M., Learner, J. (2016) Small Hive Beetle Distribution, (National Bee Unit). BBKA News Incorporating the British Bee Journal, June 2016: 204-205 pp.

Buchholz, S., Neumann, P., Merkel, K., Hepburn, H. (2006) Evaluation of *Bacillus thuringiensis* Berliner as an alternative control of small hive beetles, *Aethina tumida* Murray (Coleoptera: Nitidulidae). J. Pest Sci. 79: 251-254 pp.

Buchholz, S., Schafer, M., Spiewok, S., Pettis, J., Duncan, M., Ritter, W., Spooner-Hart, R., Neumann, P. (2008) Alternative food sources of *Aethina tumida* (Coleoptera: Nitidulidae) J. Apic. Res. Bee World 47(3): 201-208 pp.

Buchholz, S., Merkel, K., Spiewok, S., Pettis, J., Duncan, M., Spooner-Hart, R., Ulrichs, C., Ritter, W., Neumann, P. (2009) Alternative control of *Aethina tumida* Murray (Coleoptera: Nitidulidae) with lime and diatomaceous earth. Apidologie 40(5): 535-548 pp.

Cabanillas, H.E., Elzen, P. J. (2006) Infectivity of entomopathogenic nematodes (Steinernematidae and Heterorhabditidae) against the small hive beetle *Aethina tumida* (Coleoptera: Nitidulidae). J. Apic. Res. Bee World 45(1): 49-50 pp.

Calderon, R.A., Arce, H., Ramirez, J. (2006). The small hive beetle *Aethina tumida* Murray, an important problem affecting honey bees. El pequeno escarabajo de la colmena *Aethina tumida* Murray, un problema importante que afecta las abejas meliferas. Ciencius Veterinarias (Heredia), 24(1) 49-55 pp.

Calderon, Fallas, R.A., Montero, M. R., Arias, F.R., Villagra, W.V. (2015) Primer reporte de la presencia del pequeno escarabajo de la colmena *Aethina tumida*, en colmenas de abejas Africanizadas in Nicaragua. Rev. Ceinc. Vet. 32(1): 29-33 pp.

Cepero, A., Higes M., Martinez-Salvador, A., Meana A., Martin-Hernandez R. (2014) A two year national surveillance for *Aethina tumida* reflects its absence in Spain. BMC Res. Notes, 7 (878).

Cervancia, C., de Guzman, L., Polintan, E., Dupo, A., Locsin, A. (2016) Small hive beetle infestation in the Philippines. J. Apic. Res. DOI. 10.1080/00218839.2016.1194053.

Chauzat, P., Laurent, M., Kryger, P., Mutinelli, F., Roelandt, S., Roels, S., van der Stede, Y., Schafer, M., Franco, S., Duquesne, V., Riveire, M., Ribiere-Chabert, M., Hendrikx, P. (2015) Guidelines for the surveillance of the small hive beetle (*Aethina tumida*) infestation, European Union Reference Laboratory for honey bee health (EURL). Anses, 19 pp.

Conklin, T. M. (2012) Investigations of small hive beetle-yeast associations. Dissertation in partial fulfillment of PhD, Entomology Program, Penn State University, University Park, Pennsylvania, US. 148 pp.

Connor, L. (2011) The Big Island in crisis: part two of the small hive beetle story in Hawaaii. Bee Culture. 140: 23-27 pp.

Cribb, B., Rice, S., Leemon, D. (2013) Aiming for the management of the small hive beetle, *Aethina tumida*, using relative humidity and diatomaceous earth. Apidologie 44(3): 241-253 pp.

Cuthbertson, A., Mathers, J., Blackburn, L., Wakefield, M., Collins, L. (2008) Maintaining *Aethina tumida* (Coleopera: Nitidulidae) under quarantine laboratory conditions in the UK and preliminary observations on its behavior. J. Apic. Res. 47: 192-193 pp.

Cuthbertson, A.G., Brown, M. (2009) Issues affecting British honey bee biodiversity and the need for conservation of this important ecological component. International Journal of Environmental Sciences and Technology, 6(4): 695-699 pp.

Cuthbertson, A.G., Wakefield, M., Powell, M., Blackburn, L., Brown, M. (2013) The small hive beetle *Aethina tumida*: A review of its biology and control measures. Current Zoology 95(5): 644-653 pp.

Damasco, D.P. (2015) Detectan en Costa Rico escarabajo que afecta colmenas.

Darius, J.L.M. (2014) Small hive beetle infestation (*Aethina tumida*) Cuba. OEI report.

de Graff, D.C., Alippi, A., Antunez, K., Aronstein, K., Budge, G., de Koker, D., de Smet, L., Dingman, D., Evans, J., Foster, L., Funfhaus, A., Gonzalez, E., Gregorc, A., Human, H., Murray, K., Nguyen, K., Poppinga, L., Spivak, M., vanEngelsdorp, D., Wilkins, S., Gensersch, E., (2013) Standard methods for American foulbrood research. J. Apic. Res. 52(1), DOI 10.3896/IBRA.1.52.1.11.

de Guzman, L., Frake, A., Rinderer, T., Arbogast, R. (2001) Effect of Height and Color on the Efficiency of Pole Traps for *Aethina tumida* (Coleoptera; Nitidulidae) J. Econ. Entomol 104(1): 26-31 pp.

de Guzman, L., Rinderer, T., Frake, A., Tubbs, H., Elzen, P., Westervelt, D. (2006) Some observations on the small hive beetle, *Aethina tumida* Murray in Russian honey bee colonies. Am. Bee. J. 146: 618-620 pp.

de Guzman, L., Frake, A., (2007) Temperature affects *Aethina tumida* (Coleoptera: Nitidulidae) development. J. Apic. Res 46(2): 88-93 pp.

de Guzman, L., Frake, A., Rinderer, T. (2008) Detection and removal of brood infested with eggs and larvae of small hive beetles (*Aethina tumida* Murray) by Russian honey bees. J. Apic. Res. Bee World 47(3): 216-221 pp.

de Guzman, L., Prudente, J., Rinderer, T., Frake, A., Tubbs, H. (2009) Population of small hive beetles (*Aethina tumida* Murray) in two apiaries having different soil textures in Mississippi. Science of Bee Culture 1(1), 4-7 pp.

de Guzman, L., Frake, A., Rinderer, T. (2010) Seasonal population dynamics of small hive beetles, *Aethina tumida* Murray, in the south-eastern USA. J. Apic. Res. Bee World 49(2): 186-191 pp.

de Guzman, L.I., Frake, A.M., Rinderer, T.E., Arbogast, R.T. (2011) Effect of height and color on the efficiency of pole traps for *Aethina tumida* (Coleoptera: Nitidulidae). J. Econ Entomol. 104(1): 26-31 pp.

de Guzman, L., Frake, A., Rinderer, T., Pollet, D. (2017) Small Hive Beetle Leaflet. USDA/ARS Honey Bee Breeding, Genetics and Physiology Laboratory, Baton Rouge, Louisiana: 3 pp.

Delaplane, K. (1998) the small hive beetle, *Aethina tumida*. A new beekeeping pest. Athens, Georgia, USA: University of Georgia. 2 pp. http://www.bugwood.org/factsheeets/small_hive_beetle.html

Delaplane, K.S., Meyer, D.F. (2000) Crop Pollination by Bees. Wellington, UK: CAB Publishing. xv + 344 pp.

Delaplane, K.S., Ellis, J.D., Hood, W.M. (2010) A test for interactions between *Varroa destructor* (Acari: Varroidae) and *Aethina tumida* (Coleoptera: Nitidulidae) in colonies of honey bees (Hymenoptera: Apidae). Annals of the Entomological Society of America Vol 103, no. 5, 711-715 pp.

Del Valle Molina, J.A. (2007) Small hive beetle infestation (*Aethina tumida*) in Mexico: Immediate notification report. Ref. OIE: 6397, Report Date: 26/10/2007.

Diaz, W.F. (2016) Small Hive Beetle Infestation (*Aethina tumida*) Nicaragua. Salud Animal, Instituto de Proteccion y Sanidad Agropecuaria (IPSA). Managua, Nicaragua.

Dixon, D., Lafreniere, R. (2002) The small hive beetle in Manitoba. Manitoba Beekeeper, Fall 2002.

Donalson, J.M. (1989) *Oplostomus fuligineus* (Coleopteran: Scarabaeidae): Life Cycle and Biology under Laboratory Conditions, and Its Occurrence in Bee Hives. Coleopt. Bull. 43(2): 177-182 pp.

Downey, D., Chun, S., Follett, P. (2015) Radiobiology of small hive beetle (Coleoptera: Nitidulidae) and prospects for management using sterile insect releases. J. Econ. Entomol. 1-5: 868-872 pp.

Dubuc, M. (2013) Small hive beetle infestation (*Aethina tumida*), Canada. OIE report.

Duehl, A., Arbogast, R., Sheridan, A., Teal, P. (2012) The influence of light on small hive beetle (*Aethina tumida*) behavior and trap capture. Apidologie 43(4): 417-424 pp.

Eischen, F.A, Westervelt, T.D., Randall, C. (1999) Does the small hive beetle have alternate food sources? Bee Culture 139(2): 129 p.

Ellis, J.D., Neumann, P., Hepburn, H.R., Elzen, P.J. (2002a) Reproductive success and longevity of adult small hive beetles (*Aethina tumida* Murray, Coleoptera: Nitidulidae) fed different natural diets. Journal of Economic Entomology 95(5): 902-907 pp.

Ellis, J., Pirk, C., Hepburn, H., Khastberger, G., Elzen, P. (2002b) Small hive beetles survive in honey bee prisons by behavioral mimicry. Naturwissenschaften 89: 326-328 pp.

Ellis, J., Delaplane, K., Hood, W. (2002c) Small Hive Beetle (*Aethina tumida* Murray) weight, gross biometry, and sex proportion at three locations in the Southeastern United States. Am. Bee J. 142: 520-522 pp.

Ellis, J.D., Delaplane, K.S., Hepburn, H.R., Elzen, P.J. (2002d) Controlling small hive beetles (*Aethina tumida* Murray) in honey bee (*Apis mellifera*) colonies using a modified entrance. Am. Bee J. 142: 288-290 pp.

Ellis, J.D., Delaplane, K.S., Hepburn, H.R., Elzen, P.J. (2003) Efficacy of modified hive entrances and a bottom screen device for controlling *Aethina tumida* (Coleoptera: Nitidulidae) infestations in *Apis mellifera* (Hymenoptera: Apidae) colonies. J. Econ. Entomol. 96: 1647-1652 pp.

Ellis, J., Hepburn, H., Delaplane, K., Elzen, P. (2003a) A scientific note on small hive beetle (*Aethina tumida*) ovipostion and behaviour during European

(*Apis mellifera*) honey bee clustering and absconding events. J. Apic. Res. 42: 47-48 pp.

Ellis, J., Hepburn, H., Delaplane, K., Elzen, P. (2003b) the effects of adult small hive beetles, *Aethina tumida* (Coleoptera: Nitidulidae), on nests and flight activity of Cape and European honey bees (*Apis mellifera*). Apidologie 34; 399-408 pp.

Ellis, J.D. (2004) The ecology and control of small hive beetles (*Aethina tumida* Murray). PhD dissertation, Rhodes University, Grahamstown, South Africa; 385 pp.

Ellis, J., Hepburn, R., Luckman, B., Elzen, P. (2004) Effects of soil type, moisture, and density on pupation success of *Aethina tumida* (Coleoptera; Nitidulidae). Environ. Entomol. 33(4), 794-798 pp.

Ellis, J., Delaplane, K. (2007) The effects of three acaricides on the development biology of small hive beetles (*Aethina tumida*). J. Apic. Res. Bee World 46(4): 256-259 pp.

Ellis, A., Delaplane, K. (2008) Effects of nest invaders on honey bee (*Apis mellifera*) pollination efficacy. Agric. Ecosyst. Environ. 127(3-4): 201-206 pp.

Ellis, J.D., Spiewok, S., Delaplane, K.S., Buchholz, S., Neumann, P., Tedders, W.L. (2010) Susceptibility of *Aethina tumida* (Coleoptera: Nitidulidae) larvae and pupae to entomopathogenic nematodes. Journal of Economic Entomology. 103: 1-9 pp.

Ellis, J. and Ellis, A. (2016) Featured Creatures: Small Hive Beetle: *Aethina tumida*. University of Florida, Dept. of Entomology and Nematology, Gainesville, FL. Publication No. EENY474. 6 pp.

Elzen, P.J., Baxter, J.R., Westervelt, D., Randall, C., Cutts, L., Wilson, W., Eishen, F., Delaplane, K., Hopkins, D. (1999) Status of the small hive beetle in the US. Bee Culture 127(1): 28-29 pp.

Elzen, P.J., Baxter, J.R., Westervelt, D., Randall, C., Delaplane, K.S., Cutts, L., Wilson, P., Milson, W.T. (1999) Field control and biology studies of a new pest species, *Aethina tumida* Murray (Coleoptera, Nitidulidae), attacking European honey bees in the Western Hemisphere. Apidologie 30(5): 361-366 pp.

Elzen, P.J., Baxter, J.R., Neumann, P., Solbrig, A., Park, C., Hepburn, H.R., Westervelt, D. (2001) Behavior of an African and western honey bee subspecies toward the small hive beetle, *Aethina tumida*. Abstracts of the 37th International Apicultural Congress, Durban, South Africa; 40 pp.

Evans, J.D., Pettis, J.S., Shiminuki, H. (2000) Mitochondrial DNA relationships in an emergent pest of honey bees: *Aethina tumida* (Coleoptera: Nitidulidae) from the United States and Africa . Annals of the entomological Society of America 93(30): 415-420 pp.

Evans, J.D., Pettis, J.S., Hood, W.M., Shiminuki, H. (2003) Tracking an invasive honey bee pest; mitochondrial DNA variation in North American small hive beetles. Apidologie 34(4): 103-109 pp.

Evans, J.D., Spiewok, S., Teixeira, E., Neumann, P. (2008) Microsatellite loci for the small hive beetles, *Aethina tumida*, a nest parasite of honey bees. Mol. Ecol. Resour. 8(3): 698-700 pp.

Eyer, M., Chen, Y., Pettis, J.S., Newmann, P. (2008) Small hive beetle, *Aethina tumida*, is a potential biological vector of honeybee viruses. Apidologie. 40: 419-428 pp.

Eyer, M., Chen, Y., Schafer, M., Pettis, J., Neumann, P. (2009) Honey bee sacbrood virus infects adult small hive beetles, *Aethina tumida* (Coleoptera: Nitidulidae). J. Apic. Res. Bee World 48(4): 296-297 pp.

FERA (Food and Environment Research Agency) (2010) The Small Hive Beetle: a serious threat to European apiculture. Sand Hutton, UK: Food and Environment Research Agency, 23 pp.

Fletcher, M.J., Cook, L.G. (2005) Small Hive Beetle Agnote, NSW-Agriculture, New South Wales. 3 p.

Fore, T. (1998) Hive beetle still limited officially to three states. Speedy Bee 27(7): 2 pp.

Gillespie, P., Staples, J., King, C., Fletcher, M. J., Dominiak, B. C. (2003) Small hive beetle, *Aethina tumida* (Murray) (Coleoptera; Nitidulidae) in New South Wales. General and Applied Entomology 32: 5-7 pp.

Giovenazzo, P., Boucher, C. (2010) A scientific note on the occurrence of the small hive beetle (*Aethina tumida* Murray) in Southern Quebec. Am. Bee J. 150: 275-276 pp.

Greco, M., Hoffman, D., Dollin, A., Duncan, M., Spooner-Hart, R., Neumann, P. (2010) the alternative Pharaoh approach: stingless bees mummify beetle parasites alive. Naturwissenschaften 97(3): 319-323 pp.

Gutierrez, M.R. (2014) Small hive beetle infestation (*Aethina tumida*), Nicaragua. OEI report.

Habeck, D. (2002) Nitidulidae. In: Arnett, R., Thomas, M., Skelley, P., Frank, J. (eds) American Beetles, CRS Press, Boca Raton, Vol. 2 pp. 311-315.

Haque, N., Levot, G. (2005) An improved method of laboratory rearing the small hive beetle *Aethina tumida* Murray (Coleoptera: Nitidulidae), Gen. Appl. Entomol. 34: 29-30 pp.

Halcroft, M., Spooner-Hart, R., Neumann, P. (2011) Behavioral defense strategies of the stingless bee, *Austroplebeia australis*, against the small hive beetle, *Aethina tumida*. Insect. Soc. 58(2): 245-253 pp.

Hassan, A.R., Neumann, P. (2008) A survey for the small hive beetle in Egypt. J. Apic. Res. Bee World 47(3) 185-186 pp.

Hayes, R., Rice, S., Amos, B., Leemon, D. (2015) Increased attractiveness of honey bee hive product volatiles to adult small hive beetle, *Aethina tumida*, resulting from small hive beetle larval infestation. Entomol. Exp. Appl. 155: 240-248 pp.

Hepburn, H.R., Radloff, S. (1998) *Honeybees of Africa*. Springer Verlag; Berlin, Germany: 370 pp.

Hernandez, B.N. (2015) Small Hive Beetle Infestation (*Aethina tumida*) Costa Rica. Dir. General, Service Nacional de Salud Animal (SENASA) Ministerio de Agricultura y Ganderia, BARREAL DE HEREDIA, Costa Rica.

Hoffman, D., Pettis, J.S., Newmann, P. (2008) Potential host shift of the small hive beetle (*Aethina tumida*) to bumble bee colonies (*Bombus impatiens*). Insectes Sociaux. 55 (2): 153-162 pp.

Hood, W.M. (1999) Small Hive Beetle. Clemson University Entomology Information Series; Clemson, South Carolina USA; EIIS/AP-2 (revised): 4 pp.

Hood, W.M. (2000) Overview of the small hive beetle, *Aethina tumida*, in North America. Bee World 81(3): 129-137 pp.

Hood, W.M. (2004) The small hive beetle, *Aethina tumida*: a review. Bee World 85(3): 51-59 pp.

Hood, W.M. (2006) Evaluation of two small hive beetle traps in honey bee colonies. American Bee Journal 146(10): 873-876 pp.

Hood, W.M. (2009a) What integrated pest management means for today's beekeeper. Bee Culture 137: 31-37.

Hood, W.M. (2009b) Risk of feeding honey bee colonies pollen substitute patties in winter when small hive beetles, *Aethina tumida* Murray (Coleoptera: Nitidulidae) are present. Science of Bee Culture 1, No.1: 13-15 pp.

Hood, W.M. (2010) Small hive beetle IPM. Bee Culture 138: 63-65.

Hood, W.M. (2011) Handbook of Small Hive Beetle IPM. Clemson University, Cooperative Extension Service. Extension Bulletin 160: 20 pp.

Hood, W.M. (2015) Small Hive Beetle Management for the Commercial Bee-keeper. Bee Farmer 1(4): 7-9 pp.

Hood, W.M., Miller, G.A. (2005) Evaluation of an upper hive entrance for control of *Aethina tumida* (Coleoptera: Nitidulidae) in colonies of honey bees (Hymenoptera: Apidae). American Bee Journal 143(5): 405-409 pp.

Hood, W.M. and Tate B. (2010) Freeman Small Hive Beetle Trap Investigations. In Proceedings of the American Bee Research Conference, published in the American Bee Journal Vol. 150 No. 5: 502 p.

Hopkins, D., Nalepa, C., Hackney, G., Kidd, K. (1999) Studies of the small hive beetle, *Aethina tumida*, in North Carolina. American Bee Journal 139(7): 536 p.

Horridge, M., Madden, J., Witter, G. (2005) The impact of the 2002-2003 drought on Australia. J. Policy Model 27: 285-308 pp.

James, D., Faulder, R., Vogele, B., Moore, C. (2000) Pheromone-trapping of *Carpophilus* spp. (Coleoptera: Nitidulidae) in stone fruit orchards near Gosford, New South Wales: Fauna, seasonality and effect of insecticides. Aust. J. Entomol. 39: 310-315 pp.

Kanga, L., Somorin, A. (2011) Susceptibility of the small hive beetle, *Aethina tumida* (Coleoptera: Nitidulidae), to insecticides and insect growth regulators. Apidologie 43(1): 95-102 pp.

Keller, J. (2002) Testing effects of alternative diets on reproduction rates of the small hive beetle, *Aethina tumida*. MSc thesis, N.C. State University, Raleigh, NC.

Knipling, E.F. (1979) The Basic Principles of Insect Population Suppression and Management. Science and Education Administration, United States Department of Agriculture (USDA) Handbook Number 512, Issued September 1979: 659 pp.

Lawal, O., Banjo, A. (2008) Seasonal variations of pests and parasites associated with honey bees (*Apis mellifera adansonii*) in southwestern Nigeria. Acad. J. Entomol. 1(1): 1-6 pp.

Learner, J., Brown, M., Wilford, J., Flatman, I. (2015) The small hive beetle: a serious threat to European Apiculture. Animal & Plant Health Agency, UK: 27 pp.

Leemon, D. (2012) In-hive fungal biocontrol of small hive beetle. Rural Industries Research and Development Corporation Publication no. 12/012. http://rirdc.infoservices.com.au/items/12-012.

Leemon, D., McMahon, J. (2009) Feasibility study into in-hive fungal biocontrol of small hive beetle. Rural Industries Research and Development Corporation Publication no. 09/090. http://rirdc, infoservices.com. au/items/09-090.

Levot, G.W. (2008) An insecticidal refuge trap to control adult small hive beetle, *Aethina tumida* Murray (Coleoptera: Nitidulidae) in honey bee colonies. J. Apic. Res. Bee World 47(3): 222-228 pp.

Levot G.W. (2012) Commercialisation of the Small Hive Beetle Habourage Device. Rural Industries Research and Development Corporation Publication No. 11/122.

Levot, G., Somerville, D. (2012) Efficacy and safety of the insecticidal small hive beetle refuge trap APITHOR™ in bee hives. Aust. J. Entomol. 51: 198-204 pp.

Lin, H.C., Phelan, P., Bartelt, R. (1992) Synergism between synthetic food odours and the aggregation pheromone for attracting *Carpophilus lugubris* in the field. Environ. Entomol. 21: 156-159 pp.

Lounsberry, Z., Spiewok, S., Pernal, S., Sonstegard, T., Hood, W., Pettis, J., Neumann, P., Evens, J. (2010) Worldwide diaspora of *Aethina tumida* (Coleoptera; Nitidulidae), a nest parasite of honey bees. Ann. Entomol. Soc. Am. 103(4) 671-677 pp.

Loza, L.M., Alvarez, L., Ugalde, J. (2014) manual: Neuvos manejesos en la apicultura para el control del pequeno escarabajo de la colmena.

Lundie, A. E. (1940) The small hive beetle, *Aethina tumida*. South Africa Department of Agriculture & Forestry Science Bulletin No. 220; 30 pp.

Meikle, W., Patt, J. (2011) The effects of temperature, diet and other factors on development, survivorship and oviposition of *Aethina tumida* (Coleoptera: Nitidulidae). Apiculture and Social Insects 104(3): 753-763 pp.

Meikle, W., Diaz, R. (2012) Factors affecting pupation success of the small hive beetle, *Aethina tumida*. J. Insect Sci. 12(118): 1-9 pp.

Messina, F. J., Slade, A.F. (1999) Expression of a life-history trade-off in a seed beetle depends on environmental context. Physiological Entomology 24: 358-363 pp.

Michener, C.D. (2000) The Bees of the World. The Johns Hopkins University Press, Baltimore, Maryland, USA.

Milian, J.L. (2012) Reporte de notificacion de *Aethina tumida* a la OEI, Direccion del Instituto de Medicina Veterinaria, Ministerio de la Agricultura, La Habana, Cuba.

Muerrle, T., Neumann, P., Dames, J., Hepburn, H., Hill, M. (2006) Susceptibility of adult *Aethina tumida* (Coleoptera; Nitidulidae) to entomopathogenic fungi. J. Econ. Entomol. 99(1): 1-6 pp.

Mulherin, T. (2009) Stepping up the small hive beetle battle. Minister for Primary Industries, Fisheries and Rural and Regional Queensland, Queensland Government.

Mullin, C., Frazier, M., Frazier, J., Ashcraft, S., Simonds, R., vanEngelsdorf, D., Pettis, J. (2010) High Levels of Miticides and Agrochemicals in North American Apiaries: Implications for Honey Bee Health. PLoS ONE 5(3), e9754. doi:10.1371/journal.pone.0009754.

Murilhas, A.M. (2004) *Aethina tumida* arrives in Portugal. Will it be eradicated? EurBee Newslett 2: 7-9 pp.

Murray, A. (1867) List of Coleoptera received from Old Calabar, on the west coast of Africa. *The Annals and Magazine of Natural History, London* 19: 176-177 pp.

Murrle, T., Neumann, P. (2004) Mass production of small hive beetles (*Aethina tumida*, Coleoptera: Nitidulidae) J. Apic. Res 43(20): 144-145 pp.

Mustafa, S. (2015) Reproduktionsbiologie and olfaktorisches Verhalten des Kleinen Beutenkafers *Aethina tumida* Murray 1867 (Nitidulidae). PhD thesis, University of Hohenheim, Germany.

Mustafa, S., Rosenkranz, P., Tolasch, T., Steidle, H. (2006) Chemotactic orientation of the small hive beetle (*Aethina tumida*, Nitidulidae) in laboratory bioassays, in: Proceedings of the 2nd European Conference of Apidology EurBee, Prague (Czech Republic) 10-16 September 2006. (Eds. Vladimir Vesely, Marcela Vofechowska and Dalibor Titera), Bee Research Institute Dol, CZ, 47 p.

Mutinelli, F., Montarsi, F., Federico, G., Granato, A., Ponti, A., Grandinetti, G., Chauzat, M. (2014) Detection of *Aethina tumida* Murray (Coleoptera: Nitidulidae) in Italy: outbreak and early reaction measures. J. Apic. Res. 53(5): 569-575 pp.

Mutsaers, M. (2006) Beekeepers observations on the small hive beetle (*Aethina tumida*) and other pests in bee colonies in West and East Africa, in: Proceedings of the 2nd European Conference of Apidology EurBee, Prague (Czech Republic), 10-16 September 2006, (Eds. Vladimir Vesely, Marcela Vorechovská and Dalibor Titera), Bee Research Institute Dol, CZ, 44 pp.

Nasr, M.E. (2001) Methods of sampling and measuring populations of tracheal mites. In: Mites of the Honey Bee, eds. Webster, T. and Delaplane, K., Published by Dadant & Sons, Inc., Hamilton, Illinois, USA, 80 p.

Neumann, P., Pirk, C., Hepburn, H., Solbrig, A., Ratnieks, F., Elzen, P., Baxter, J. (2001) Social encapsulation of beetle parasites by Cape honey bee colonies (*Apis mellifera capensis* Esch.). Naturwissenschaften 88: 214-216 pp.

Neumann, P., Hartel, S. (2004) Removal of small hive beetle (*Aethina tumida*) eggs and larvae by African honey bee colonies (*Apis mellifera scutellata*). Apidologie 35: 31-36 pp.

Neumann, P., Ellis, J.D. (2008) The small hive beetle (*Aethina tumida* Murray, Coleoptera: Nitidulidae): distribution, biology, and control of an invasive species. J. Apic. Res. Bee World. 47(3): 180-183 pp.

Neumann, P., Elzen, P. (2004) The biology of the small hive beetle (*Aethina tumida*, Coleoptera: Nitidulidae): Gaps in our knowledge of an invasive species. Apidologie 35: 229-247 pp.

Neumann, P., Hoffman, D., Pettis, J.S. (2008) Potential host shift of the small hive beetle (*Aethina tumida* Murray) to bumblebee colonies (*Bombus impatiens* Cresson). Insectes Sociaux. 55: 153-162 pp.

Neumann, P., Hoffman, D., Duncan, M., Spooner-Hart, R. (2010) High and rapid infestation of isolated commercial honey bee colonies with small hive beetles in Australia. J. Apic. Res. Bee World 49(4). 343-344 pp.

Neumann, P., Hoffman, D., Duncan, M., Spooner-Hart, R., Pettis, J. (2012) Long-range dispersal of small hive beetles. J. Apic. Res. 51(2): 214-215 pp.

Neumann, P., Evans, J., Pettis, J., Pirk, C., Schaffer, M., Tanner, G., Ellis, J. (2013) Standard methods for small hive beetle research. J. Apic. Res. 52(4): 1-32 pp.

Neumann, P., Naef, J., Crailsheim, K., Crewe, R., Pirk, C. (2015) Hit and run trophallaxis of small hive beetles. Ecol. Evol. Doi: 10.1002/ece3.1806/full.

Neumann, P., Buchholz, S., Jenkins, M., Pettis, J. (2016a) The suitability of sterile insect technique as a pest management of small hive beetles, *Aethina tumida* Murray (Coleoptera: Nitidulidae). J. Apic. Res. 54(3): 236-237 pp.

Neumann, P., Pettis, J., Schafer, M. (2016b) Quo vadis *Aethina tumida* ? Biology and control of small hive beetles. Apidologie 47: 427-466 pp.

Nolan, M.P. (2008) Trapping Small Hive Beetles, *Aethina tumida* Murray, Inside Honey Bee Colonies. MSc Thesis, Entomology, Clemson University, Clemson, South Carolina, USA: 54 pp.

Nolan IV, M.P., Hood, W.M. (2008) Comparison of two attractants to small hive beetles, *Aethina tumida*, in honey bee colonies. J. Apic. Res. Bee World 47(3): 229-233 pp.

Nolan, M.P., Hood, W.M. (2010) Trapping small hive beetles, *Aethina tumida*, in honey supers and brood chambers of honey bee colonies. Science of Bee Culture: Vol. 2, No.1, pp 8-11 pp.

OIE (World Organisation for Animal Health) (2012) Terrestrial Animal Health Code, edition 21. Paris, France: Office International des Epizooties. http://www.oie.int/international-standard-setting/terrestrial-code/access-online/

Palmeri,V., Scrito, G., Malacrino, A., Laudani, F., Campolo, O. (2015) A scientific note on a new pest for European honey bees: first report of *Aethina tumida* (Coleoptera: Nitidulidae) in Italy. Apidologie 46(4): 527-529 pp.

Park, A.L., Pettis, J.S., Caron, D.M. (2002) Use of household products in control of small hive beetle larvae and salvage of treated combs. American Bee Journal 142(6): 439-442 pp.

Pena, W.L., Carballo, L., Lerenzo, J. (2014) Reporte de *Aethina tumida* Murray (Coleoptera, Nitidulidae) en colonias de la abeja sin aguijon *Melipona beecheii* Bennett de Matanzas y Mayabeque. Rev. Salud. Anim. 36(3): 201-204 pp.

Peterson, S. (2012) Trapping and Control of the Small Hive Beetle, *Aethina tumida*, an Invasive Parasite of Honey Bees, *Apis mellifera*. MSc. Thesis (Entomology), Clemson University, Clemson, South Carolina, USA. 68 pp.

Petroski, R., Bartelt, R., Vetter, R. (1994) Male produced aggregation pheromone of *Carpophilus obsoletus* (Coleoptera, Nitidulidae). J. Chem. Ecol. 20: 1483-1493 pp.

Pettis, J., Shimanuki, H. (2000) Observations on the small hive beetle, *Aethina tumida* Murray, in the United States. American Bee Journal 140(2): 152-155 pp.

Pettis, J., Martin, D., vanEngelsdorp, D. (2014) Migratory Beekeeping. In: Ritter, W. (ed.) Bee Health and Veterinarians, pp. 51-54. OIE, Paris.

Pirk, C., Neumann, P. (2013) Small hive beetles are facultative predators of adult honey bees. J. Insect Behav. 26: 796-803 pp.

Quigley, A. (2015a) Scorched Earth in Italy: Italy and Small Hive Beetles. Bee Culture: www.beeculture.com/scorched-earth-italy.

Quigley, A. (2015b) Nervous Neighbors: Small Hive Beetle in Italy, And Maybe Beyond! Bee Culture: www.beeculture.com/nervous-neighbors.

Richards, C., Hill, M., Dames, J. (2005) The susceptibility of small hive beetle (*Aethina tumida* Murray) pupae to *Aspergillus niger* (van Tieghem) and *A. flavus* (Link: Grey). Am. Bee J. 145: 748-751 pp.

Robson, J. D. (2012) Small Hive Beetle *Aethina tumida* Murray (Coleoptera: Nitidulidae), Pest Alert 12-01. PLANT PEST CONTROL BRANCH, Division of Plant Industry, Department of Agriculture. First Issued January 2012, 1428 South King Street, Honolulu, Hawaii 96814.

Rhodes, J., McCorkell, B. (2007) Small hive beetle *Aethina tumida* in New South Wales apiaries 2002-6: survey results 2006. New South Wales Department of Primary Industry. 1-32 pp.

Sanford, M. (1998) *Aethina tumida*: a new beehive pest in the Western Hemisphere. Apis 16: 1-5 pp.

Sanford, M. (1999) Small hive beetle update. Bee Culture 127(2): 56. p.

Santino, I., Bono, S., Borruso, L., Bove. M., Cialdi, E., Martinelli, D., Alari, A. (2013) *Kodamaea ohmerei* isolate from two immunocompromised patients: first report in Italy. Mycoses 56: 179-181 pp.

Schafer, M., Pettis, J., Ritter, W., Neumann, P. (2008) A simple method of quantitative diagnosis of small hive beetles, *Aethina tumida*, in the field. Apidologie. 39: 564-565 pp.

Schafer, M., Ritter, W., Pettis, J., Teal, P., Neumann P. (2009) Effects of organic acid treatments on small hive beetles, *Aethina tumida*, and the associated yeast *Kodamaea ohmeri*. J. Pest Sci. 82(3): 283-287 pp.

Schafer, M., Ritter, W., Pettis, J., Neumann, P. (2010a) Winter losses of honey bee colonies (Hymenoptera: Apidae): The role of infestations with *Aethina tumida* (Coleoptera: Nitidulidae) and *Varroa destructor* (Parasitiformes: Varroidae). J. Econ. Entomol 103(1) 10-16 pp.

Schafer, M., Ritter, W., Pettis, J., Neumann, P. (2010b) Small hive beetles, *Aethina tumida*, are vectors of *Paenibacillus larvae*. Apidologie 41(1): 14-20 pp.

Schafer, M., Pettis, J., Ritter, W., Neumann, P. (2010c) Simple small hive beetle diagnosis. Am. Bee J. 150: 371-372 pp.

Schafer M., Ritter, W. (2014) The small hive beetle (*Aethina tumida*). In: Ritter, W. (ed) Bee health and veterinarians. World Organisation for Animal Health, Paris: 149-156 pp.

Schmolke, M.D. (1974) A study of *Aethina tumida*: the small hive beetle. MSc Thesis, University of Rhodesia, South Africa; 181 pp.

Shapiro-Ilan, D., Morales-Ramos, J., Rojas, M., Tedders, W. (2010) Effects of novel entomopathogenic nematode-infected host formulation on cadaver integrity, nematode yield, and suppression of *Diaprepes abbreviatus* and *Aethina tumida*. Journal of Invertebrate Pathology. 103: 103-108 pp.

Smith, H. (2012) Jamaican Beekeeping Industry Status. 1 p. http://hughsmithja. weebly.com/jamaica-beekeeping-industry.html

Solbrig, A.J. (2001) Interactions between the South African honey bee *Apis mellifera* capensis Esch. and the small hive beetle *Aethina tumida*. MSc. Thesis FU Berlin.

Somerville, D. (2003) Small hive beetles in the USA. A report for the Rural Industries Research & Development Corporation. Pub. No. 03/050: 57 pp.

Spiewok, S., Neumann, P. (2006a) Infestation of commercial bumblebee (*Bombus impatiens*) field colonies by small hive beetles (*Aethina tumida*). Ecological Entomology 31: 623-628 pp.

Spiewok, S., Neumann, P. (2006b) Cryptic low-level reproduction of small hive beetles in honey bee colonies. J. Apic. Res. 45(1) 47-48 pp.

Spiewok, S., Neumann, P. (2012) Sex ratio and dispersal of small hive beetles. J. Apic. Res. 51(2): 216-217 pp.

Spiewok, S., Pettis, J., Duncan, M., Spooner-Hart, R., Westervelt, D., Neumann, P. (2007) Small hive beetle, *Aethina tumida*, populations I: Infestation levels of honey bee colonies, apiaries and regions. Apidologie 38(6): 595-605 pp.

Spooner-Hart, R., Pettis, J., Duncan, M. (2016) The small hive beetle in Australia. In: Carreck, N.L. (ed.) The small hive beetle in Europe. International Bee Research Association, Groombridge.

Stedman, M. (2006) Small Hive Beetle: *Aethina tumida* Murray (Coleoptera: Nitidulidae). Primary Industries and Resources for South Australia. Factsheet 03/06: 13 pp.

Strauss, U., Human, H., Crew, R., Pirk, C. (2010) The first report of storage mites *Caloglyphus hughesi* (Acaridae) on laboratory-reared *Aethina tumida* Murray (Coleoptera: Nitidulidae) in South Africa. Afr. Entomol. 18(2): 379-382 pp.

Strauss, U., Human, H., Gauthier, L., Crew, R., Dietemann, V., Pirk, C. (2013) Seasonal prevalence of pathogens and parasites in the savannah honey bee (*Apis mellifera scutellata*). J. Invertebr. Pathol. 114(1): 45-52 pp.

Suazo, A., Torto, B., Teal, P.E., Tumlinson, J.H. (2003) Response of the small hive beetle (*Aethina tumida*) to honey bee (*Apis mellifera*) and beehive-produced volatiles. Apidologie 34(6): 525-533 pp.

Teixeira, E.W., De Jong, D., Sattler, A., Message, M. (2016) *Aethina tumida* Murray (Coleoptera, Nitidulidae), a pequena besouro das colmeias, chega ao Brasil. Mensagem Doce 136: 2-10 pp.

Tew, J. E. (2015) Beeyard Thoughts: Microfiber towels and small hive beetles. Bee Culture, November 20, 2015 (www.Beeculture.beeyard-thoughts-3).

Torto, B., Arbogast, R., van Engelsdorp, D., Williams, S., Purcell, D., Boucias, D., Tumlinson, J., Teal, P. (2007) Trapping of *Aethina tumida* Murray (Coleoptera: Nitidulidae) from *Apis mellifera* L. (Hymenoptera: Apidae) colonies with an in-hive baited trap. Environ. Entomol. 36(5): 1018-1024 pp.

Torto, B., Fombong, A., Arbogast, R., Teal, P. (2010) Monitoring *Aethina tumida* (Coleoptera: Nitidulidae) with baited bottom board traps: occurrence and seasonal abundance in honey bee colonies in Kenya. Environ. Entomol. 39(6); 1731-1736 pp.

Toufailia, H.A., Alves, D., Bena, D., Bento, J., Iwanicki, N., Cline, A., Ellis, J., Ratnieks, L. (2017) First record of small hive beetle, *Aethina tumida* Murray, in South America. J. Apic. Res. 56(1), 76-80 pp.

The Australian Beekeeper. (2002) Restriction on small hive beetles lifted. Australian Post Publication No. NAC 1202: 104(6): 225-227.

Valerio da Silva, M.J. (2014) The first report of *Aethina tumida* in the European Union, Portugal 2004. Bee World. 91(4): 90-91 pp.

Van Engelsdorp, D., Underwood, R., Caron, D., Hayes, J. (2007) An estimate of managed colony losses in the winter of 2006-2007: a report commissioned by the Apiary Inspectors of America. Am. Bee J. 147: 599-603 pp.

Villa, J.D. (2004) Swarming behavior of honey bees (Hymenoptera: Apidae) in Southeastern Louisiana Ann. Entomol. Soc. Am. 97(1) 111-116 pp.

Ward, L., Brown, M., Neumann, P., Wilkins, S., Pettis, J., Boonham, N. (2007) A DNA method for screening hive debris for the presence of small hive beetle (*Aethina tumida*). Apidologie 38(30): 272-280 pp.

Westervelt, D. (2005) Small hive beetles in the USA –What we've learned in nine years. Am. Bee J. 145(10): 805-807 pp.

Westervelt, D., Causey, D., Neumann, P., Ellis, J., Hepburn, H. (2001) Grease patties worsen small hive beetle infestations. Am. Bee J. 141: 775 p.

White, B. (2003) Small hive beetle update. Australian Beekeeper. May 2003, No. 11: 2 pp.

Wikipedia (2017) Small Hive Beetle. http://en.wikipedia.org/wiki/Small_hive_beetle

Williams, L. (2015) The Small Hive Beetle – A Serious Threat to European Apiculture. Department for Environment, Food & Rural Affairs; Animal & Plant Health Agency Leaflet: 27 pp.

Zacchetti, F. (2015) *Aethina tumida* dal vivio. L'Apis. 2: 5-8 pp.

Zawislak, J. (2014) Managing Small Hive Beetles. University of Arkansas. Division of Agriculture, Agriculture and Natural Resources Bulletin No. FSA7075: 1-6 pp.

This information is supplied with the understanding that no discrimination is intended and no endorsement by the author is implied. Brand names of pesticides are given as a convenience and are neither an endorsement nor guarantee of the product nor a suggestion that similar products are not effective. Use pesticides only according to the directions on the label. Follow all directions, precautions and restrictions that are listed.

List of figures and corresponding page numbers

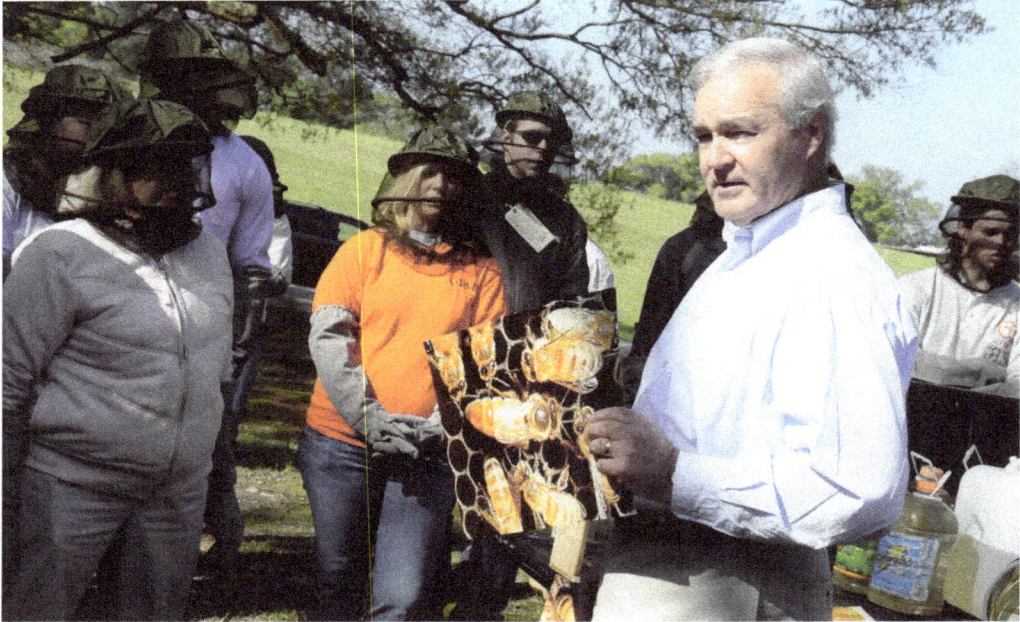

About the author

The author is Professor Emeritus of Entomology, Emeritus College, Clemson University, South Carolina, USA where he retired in 2013. He began work at Clemson University in 1988 with a split appointment as Extension Honey Bee Specialist in the College of Agriculture and as State Apiarist in the Department of Plant Industry in the Division of Regulatory Services. In 1995, he began duties as State Apiculturist in the Department of Entomology where he taught undergraduate apiculture, served as the State Extension Honey Bee Specialist and conducted honey bee research specializing in honey bee pest management including research on honey bee tracheal mites, varroa mites, wax moths, and small hive beetles. His research for the last 15 years at the university focused primarily on small hive beetle integrated pest management beginning in 1998 on Wadmalaw Island, Charleston County, South Carolina where he began development of a small hive beetle trap, later known as the "Hood Beetle Trap." He is also retired member of the US Army serving two years on active duty and serving 23 more years in the active reserves. He retired from service at the rank of Lieutenant Colonel in 2011. His hobbies in retirement include fishing, hunting, golf, travel and spending time with his nine grandchildren.

www.ingramcontent.com/pod-product-compliance
Lightning Source LLC
Chambersburg PA
CBHW081507200326
41518CB00015B/2411